Springer Theses

Recognizing Outstanding Ph.D. Research

Aims and Scope

The series "Springer Theses" brings together a selection of the very best Ph.D. theses from around the world and across the physical sciences. Nominated and endorsed by two recognized specialists, each published volume has been selected for its scientific excellence and the high impact of its contents for the pertinent field of research. For greater accessibility to non-specialists, the published versions include an extended introduction, as well as a foreword by the student's supervisor explaining the special relevance of the work for the field. As a whole, the series will provide a valuable resource both for newcomers to the research fields described, and for other scientists seeking detailed background information on special questions. Finally, it provides an accredited documentation of the valuable contributions made by today's younger generation of scientists.

Theses are accepted into the series by invited nomination only and must fulfill all of the following criteria

- They must be written in good English.
- The topic should fall within the confines of Chemistry, Physics, Earth Sciences, Engineering and related interdisciplinary fields such as Materials, Nanoscience, Chemical Engineering, Complex Systems and Biophysics.
- The work reported in the thesis must represent a significant scientific advance.
- If the thesis includes previously published material, permission to reproduce this must be gained from the respective copyright holder.
- They must have been examined and passed during the 12 months prior to nomination.
- Each thesis should include a foreword by the supervisor outlining the significance of its content.
- The theses should have a clearly defined structure including an introduction accessible to scientists not expert in that particular field.

More information about this series at http://www.springer.com/series/8790

Hamidreza Arandiyan

Methane Combustion over Lanthanum-based Perovskite Mixed Oxides

Doctoral Thesis accepted by
Tsinghua University, Beijing, China

Author
Dr. Hamidreza Arandiyan
School of Chemical Engineering
The University of New South Wales
Sydney
Australia

Supervisor
Prof. Junhua Li
School of Environment
Tsinghua University
Beijing
China

ISSN 2190-5053 ISSN 2190-5061 (electronic)
Springer Theses
ISBN 978-3-662-51666-9 ISBN 978-3-662-46991-0 (eBook)
DOI 10.1007/978-3-662-46991-0

Springer Heidelberg New York Dordrecht London

Softcover reprint of the hardcover 1st edition 2015

Printed on acid-free paper

Springer-Verlag GmbH Berlin Heidelberg is part of Springer Science+Business Media
(www.springer.com)

Parts of this thesis have been published in the following journal articles:

Scholarly Book Chapter

1. **Arandiyan, H.**, Li, J., 2012, Catalytic CO2 reforming of methane over perovskite noble metals, in Li, H; Xu, QJ; Zhang, D, Advanced Materials Research, pp. 1070–1074 Trans Tech Publications LTD, Laublsrutistr 24, CH-8717 Stafa, Zurich, Switzerland.

Refereed Journal Articles

2. **Arandiyan, H.**; Dai, H.; Li, J.; et al., *Pt Nanoparticles Embedded in Colloidal Crystal Templating Derived 3D Ordered Macroporous Ce0.6Zr0.3Y0.1O2: Highly Efficient Catalysts for Methane Combustion.* ACS Catalysis, 2015, 5, 1781–1793. (IF: 7.572)
3. **Arandiyan, H.**; Dai, H.; Li, J. et al., *Enhanced catalytic efficiency of Pt nanoparticles supported on three-dimensionally ordered macro-/mesoporous Ce0.6Zr0.3Y0.1O2 for methane combustion.* Small, *DOI: 10.1002/smll.201402951* (IF: 7.823) *Journal Frontispiece Cover Page.*
4. **Arandiyan, H.**; Peng, Y.; Li, J., et al., *Effects of noble metals doped on mesoporous LaAlNi mixed oxide catalyst and identification of carbon deposit for reforming CH4with CO2.* Journal of Chemical Technology & Biotechnology 2014, 89, 372–381. (IF: 2.494)
5. **Arandiyan, H.**; Dai, H.; Li, J., et al., *Three-Dimensionally Ordered Macroporous La0.6Sr0.4MnO3Supported Ag Nanoparticles for the Combustion of Methane.* Journal of Physical Chemistry C 2014, 118, 14913–14928. (IF: 4.835) *Journal Front Cover Page.*
6. **Arandiyan, H.**; Chang, H.; Li, J., et al., *Dextrose-aided hydrothermal preparation with large surface area on 1D single-crystalline perovskite La0.5Sr0.5CoO3 nanowires without template: Highly catalytic activity for methane combustion.* Journal of Molecular Catalysis A: Chemical 2013, 378, 299–306. (IF: 3.679)
7. **Arandiyan, H.**; Dai, H.; Li, J., et al., *Three-dimensionally ordered macroporous La0.6Sr0.4MnO3 with high surface areas: Active catalysts for the combustion of methane.* Journal of Catalysis 2013, 307, 327–339. (IF: 6.423)
8. **Arandiyan, H.**; Dai, H.; Li, J., et al., *Dual-templating synthesis of three-dimensionally ordered macroporous La0.6Sr0.4MnO3-supported Ag nanoparticles: controllable alignments and super performance for the catalytic combustion of methane.* Chemical Communications 2013, 49, 10748–10750. (IF: 6.718)
9. **Arandiyan, H.**; Li, J.; Ma, L.; et al., *Methane reforming to syngas over LaNixFe1−xO3 (0≤x≤1) mixed-oxide perovskites in the presence of CO2and O2.* Journal of Industrial and Engineering Chemistry 2012, 18, 2103–2114. (IF: 2.011)

Conference Submissions

10. **Arandiyan, H.**; Dai, H.; Li, J.; Amal, R.; *Highly ordered Pt/3DOM Ce0.6Zr0.3Y0.1O2 with crystalline framework for the combustion of methane.* Energy Future Conference, Australia, 2014.
11. **Arandiyan, H.**; Dai, H.; Li, J. *3DOM La0.6Sr0.4MnO3-supported Ag NPs: catalytic performance for methane combustion. 17th National Conference on Catalysis of China*, China, 2014.
12. **Arandiyan, H.**; Li, J. *Influence of the carbon on catalytic activity of noble metal perovskite-type mixed oxide for CO2reforming of methane. 15th International Congress on Catalysis,* Germany, 2012.
13. **Arandiyan, H.**; Li, J. *Template-free synthesis of single-crystalline perovskite LaSrMn nanowires: highly catalytic activity for methane combustion. 23th North American Catalyst Meeting,* USA, 2013.
14. **Arandiyan, H.**; Li, J. Dai, H.; *Ag NPs embedded in 3DOM La0.6Sr0.4MnO3 derived from the colloidal crystal templating strategy. 6th Asia-Pacific Congress on Catalysis,* Taiwan, 2013.
15. **Arandiyan, H.**; Dai, H.; Li, J. *DMOTEG-PMMA dual-templating generation of 3DOM LSMO supported silver nanoparticle. 6th Japan-China Workshop on Environmental Catalysis*, Japan, 2013.
16. **Arandiyan, H.**; Liu, C., *Enhancement on activity over novel CeO2supported on TiO2-SiO2binary metal oxide for NH3-SCR of NO. 15th International Congress on Catalysis,* Germany, 2012.
17. **Arandiyan, H.**; Liu, C. *Characterization of CeO2-WO3catalysts for selective catalytic reduction of NOx with NH3. International Conference on Environment Pollution Control,* China, 2011.
18. **Arandiyan, H.**; Li, J., *Dry reforming of methane over LaNixFe(1-x)O3 mixed-oxide perovskites as catalysts precursors. 22nd North American Catalysis Society Meeting*, USA, 2011.
19. **Arandiyan, H.**; Li, J. *Structural features of perovskites as catalysts precursors for CO2reforming of CH4on La0.4Al0.2Ni0.8M0.6O3. 5th China-Japan Environmental Catalysis,* China, 2011.

To my family for their endless love and constant support

This book is dedicated to the spirit of my dear brother, Ali Arandiyan, and in my unforgotten memories of him.

Supervisor's Foreword

As a 100-year global warming potential greenhouse gas, methane is much more effective greenhouse gas than CO_2. Natural gas is one of the most multi-purpose alternative fuels available and can be used in spark ignited or compression engines from the smallest of motorbikes to the largest rail locomotives. Catalytic oxidation of CH_4 is regarded as an attractive alternative to conventional thermal combustion for energy production. Meanwhile, three-dimensionally ordered macroporous (3DOM) and nanostructured catalysts are more and more important for highly developed applications related to photonic crystals, catalysis, and membrane separation. Recent theoretical and experimental studies have recommended that noble metals supported on perovskite-type oxides (ABO_3) with well-ordered porous networks show promising redox properties. 3DOM structure-supported noble metals have shown great potential for broad applications due to their interesting properties (e.g., high BET and ordered mesostructures).

The book presented precisely focus on an innovative predesign of a new type of perovskite (ABO_3) catalyst with excellent oxygen mobility, crystal structure, thermochemical stability, and high performance in terms of redox properties, as well as optimization of the reaction system. A high CH_4 oxidation activity utilized the available energy of CH_4 at low reaction temperatures, increasing system catalytic activity and controlling emissions by significantly reducing the required temperatures. Developed functionalized nanohybrid three-dimensionally ordered macro perovskites (3DOM ABO_3) and transition metal nanoparticles for CH_4 oxidation reaction. Additionally, the developed new strategies for synthesizing 3DOM ABO_3 nanohybrid catalysts have minimal open space with maximal surface area pore volume and a narrow pore size distribution so as to promote a lower reaction temperature.

This thesis has been highly praised by the eight Ph.D. thesis reviewers, including two academicians of Chinese Academy of Sciences. It was concluded that this thesis "gives originally analytical and theoretical results," "has both important academic significance and application value," and "is a rarely high-quality doctoral dissertation." I sincerely hope that this book will be of great help and inspiration for those young scientists who willingly devote themselves to environmental and chemical engineering, chemistry, and material science, especially to catalysis material researchers.

Beijing, China
March 2015

Prof. Junhua Li

Abstract

Three-dimensionally ordered macroporous (3DOM) and nanostructured catalysts are more and more important for highly developed applications related to photonic crystals, catalysis, and membrane separation. Recent theoretical and experimental studies have recommended that noble metals supported on perovskite-type oxides (ABO_3) with well-ordered porous networks show promising redox properties. 3DOM structure-supported noble metals have shown great potential for broad applications due to their interesting properties (e.g., high BET and ordered meso-structures). Although the macroporous metals (voids >50 nm) have been widely achieved, the fabrication of ordered macroporous perovskites with mesoporous skeletons is still challenging. The main innovations results are as follows:

In this work, 1D single-crystalline nanowires of perovskite-type oxides (PTOs) $La_{0.5}Sr_{0.5}CoO_3$ were firstly prepared by a dextrose-assisted hydrothermal route (DHR) and/or L-Lysine. It is found that the single-crystalline sample (LSCO-2) derived at temperature of 170 °C with a dextrose/L-Lysine volumetric ratio of 0.3/1.0 of new nanowire DHR possessed the high surface O_2 concentration, the best low-temperature reducibility, and the high BET surface area (17.7 $m^2\ g^{-1}$), which exhibited much better than that conventional process for polycrystalline $La_{0.5}Sr_{0.5}CoO_3$ catalyst. After running at 800 °C for 50 h, the nanowire DHR (LSCO-2) showed a higher stability and activity than the nanoparticles (LSCO-4) counterpart. It is believed that the good catalytic performance of LSCO-2 was related to factors such as higher surface area/energies and better low-temperature reducibility, as well as to the different adsorbed O_2 species concentration.

In addition, three-dimensionally ordered macroporous rhombohedral $La_{0.6}Sr_{0.4}MnO_3$ (3DOM LSMO) with nanovoids was synthesized using polymethyl methacrylate (PMMA) microspheres as hard template and dimethoxytetraethylene glycol (DMOTEG), ethylene glycol, polyethylene glycol (PEG400), L-lysine, or triblock copolymer (Pluronic P-123) as surfactant. It is found that LSMO-DP3 sample derived with 3.0 ml of DMOTEG and 5.0 ml of PEG400 possessed the higher O_2 species concentration and BET and better low-temperature reducibility, hence exhibiting a good catalytic activity for combustion of CH_4. The apparent activation energies of the 3DOM LSMO samples were estimated to be 56.5–75.2 kJ/mol, with the LSMO-DP3 sample showing the lowest apparent activation energy (56.6 kJ/mol).

Highly dispersed Ag NPs supported on high-surface-area 3DOM $La_{0.6}Sr_{0.4}MnO_3$ (yAg/3DOM $La_{0.6}Sr_{0.4}MnO_3$; $y = 0$, 1.57, 3.63, and 5.71 wt%) were successfully generated with high surface areas (38.2–42.7 m^2/g) via the dimethoxytetraethylene glycol-assisted gas bubbling reduction method. Meanwhile, the effects of H_2O and SO_2 on the catalytic activity of the 3.63 wt% Ag/3DOM $La_{0.6}Sr_{0.4}MnO_3$ sample were also examined. It is concluded that its super catalytic activity was associated with its highest oxygen adspecies concentration, good low-temperature reducibility, larger surface area, and strong interaction between metal and support as well as unique nanovoid-walled 3DOM structure.

Keywords Three-dimensionally ordered macroporous · Perovskite-type oxide · Templating preparation method · Low-temperature reducibility · Methane combustion

Preface

Three-dimensionally ordered macroporous (3DOM) and nanostructured catalysts are more and more important for highly developed applications relating in photonic crystals, catalysis, and membrane separation. Recent theoretical and experimental studies have recommended that noble metals supported on perovskite-type oxides (ABO_3) with well-ordered porous networks show promising redox properties. 3DOM structure-supported noble metals have shown great potential for broad applications due to their interesting properties (e.g., high BET and ordered meso-structures). Although the macroporous metals (voids >50 nm) have been widely achieved, the fabrication of ordered macroporous perovskites with mesoporous skeletons is still challenging. This book contains 5 chapters. The introduction (Chap. 1) covers the development of natural gas vehicle, research progress of catalytic CH_4 combustion catalyst, and viewpoint of the synthesis for perovskite-mixed oxide (ABO_3). Chapter 2 summarizes the preparation of 1D single-crystalline nanowire of perovskite-type oxides $La_{0.5}Sr_{0.5}CoO_3$, which were prepared by a dextrose-assisted hydrothermal route (DHR) and/or L-Lysine. Study on 1 Dimensional single-crystalline LSCO nanowires and their characterization such as XRD, HRSEM, HRTEM, and BET, as well as activity evaluation of catalyst such as influence of preparation method, stability, calcinations, and space gas velocity on catalytic activity. Chapter 3 is devoted to the three-dimensionally ordered macroporous rhombohedral $La_{0.6}Sr_{0.4}MnO_3$ (3DOM LSMO) with nanovoids, which were synthesized using polymethyl methacrylate (PMMA) microspheres as hard template and dimethoxytetraethylene glycol (DMOTEG), ethylene glycol, polyethylene glycol (PEG400), L-lysine, or triblock copolymer (Pluronic P-123) as surfactant. In Chap. 4, we focus on the synthesis of the highly dispersed Ag NPs supported on high-surface-area 3DOM $La_{0.6}Sr_{0.4}MnO_3$ (yAg/3DOM $La_{0.6}Sr_{0.4}MnO_3$; y = 0, 1.57, 3.63, and 5.71 wt%) were successfully generated with high surface areas (38.2–42.7 m^2/g) via the dimethoxytetraethylene glycol-assisted gas bubbling reduction method. Meanwhile, the effects of H_2O and SO_2 on the catalytic activity of the 3.63 wt% Ag/3DOM $La_{0.6}Sr_{0.4}MnO_3$ sample were also examined. In the meantime, describe the most widely used experimental techniques on the structures characterization of 3DOM materials. It is concluded that its super catalytic activity was associated with its highest oxygen adspecies concentration, good low-temperature reducibility, larger surface area, and strong interaction between metal and support as well as unique nanovoid-walled 3DOM structure. It is apparent that the 3DOM material filed is eager for more and more researchers from other field to explore attractive applications.

This book condenses the authors' great efforts and contributions. We hope that this book can provide beneficial help and inspiration for undergraduate, graduate students, scientists, and researchers who are majoring in environment, chemical engineering, chemistry, and materials. Due to the relatively wide areas covered in this book, the numerous contents with connection to complex scientific issues, together with the limited knowledge and ability of the authors, we sincerely appreciate the criticism and comments from readers.

April 2015 Dr. Hamidreza Arandiyan

Acknowledgments

First and foremost, I would like to thank Prof. Li Junhua for his guidance, support, and friendship. His enthusiasm for my research as well as his scientific support and the friendly discussions was essential in my work.

In my three years as a member of the State Key Joint Laboratory of Environment Simulation and Pollution Control (SKLESPC), I have had the opportunity to collaborate with very talented people at School of Environment, Tsinghua University. I would like to especially acknowledge my classmates for their continuous support and friendship.

I would like to acknowledge Prof. Dai Hongxing at College of Environmental and Energy Engineering, Beijing University of Technology, for the chance he gave me to work on this very interesting theme in his group at Department of Chemistry and Chemical Engineering, and his help and advices through the different steps to success. I would like to thank also Prof. Deng Jiguang at BJUT, for his strong scientific support and the friendly discussions. I would also like to thank those graduate students in Laboratory of Catalysis Chemistry and Nanoscience at Beijing University of Technology who helped build the 3DOM architecture from the group up, and got me started on the right track. They are Dr. Liu Yuxi, Li Xinwei, Dr. Xie Shaohua, Han Wen, and Dr. Zhao Zhenxuan. I owe a special thanks to Wang Yuan (Helena) for her support and advice, which helped me improving the quality of this project. Without their hard work, enthusiasm, and creativity, this work could not have been completed.

I would like to acknowledge Prof. Li Junhua's group members for their technical help in various projects. All the other members of the (SKLESPC) have contributed in one way or another to the development of this research, and for this, I thank them profoundly. They are Dr. Chang Huazhen, Dr. Chen Liang, Dr. Liu Xiacia, Dr. Ma Lei, Dr. Chen Jinghuan, Dr. Peng Yue, Bai Bingyang, Wang Chizhong, Dai Yun, Khalid, and Li Dongfang. Financial support from Collaborative Innovation Center for Regional Environmental Quality, National Natural Science Foundation of China (Grant Nos. 51078203, 21077007 and 21221004), and National High-Tech Research and Development (863) Program of China (Grant Nos. 2009AA064806, 2012AA062506 and 2013AA065304) is appreciated.

I would like to thank my good Iranian friends, Mr. Bagheri, Mr. Lotfi, Mr. Moazami, Mr. Farmand, Mr. Zahedi, Mr. Naeini, Mr. Vafaei, Majid, and Masoud Shamaizadeh, and other Iranian people who live in China with their lovely families for their unconditional support and loyalty.

My parents, Reza and Mina, my brother Mohammad, and my sister Samira have always supported me and showed me their unending care. My deepest gratitude goes to my family for their love; patience and care were the motivation and inspiration that allowed me to finish this journey. Thanks God for allowing me to take one more step in my life.

Contents

About the Author

Dr. Hamidreza Arandiyan was awarded his Ph.D. in July 2014, with a ranking of excellent from the School of Environment at Tsinghua University (THU), a Top University in China. During his Ph.D., he received several awards and fellowships working under the supervision of Prof. Junhua Li (a world leader in environmental catalysis). He was invited to do research at the Zeolite Research Group (2008) in the laboratories of Prof. H. Kazemian where he was a senior researcher in the Science and Technology Park of Tehran University. Afterward, he worked in the laboratory of Prof. Hongxing Dai as part of the Catalysis Chemistry and Nanoscience group at Beijing University of Technology, during his Ph.D. program (2013). In July 2014, just before completing his Ph.D. degree at THU, he was offered a position as a full-time research associate in the School of Chemical Engineering at the University of New South Wales (UNSW) under the supervision of *Scientia* Prof. Rose Amal. He was recently awarded (2015) a UNSW Vice-Chancellor's Postdoctoral Research Fellow.

His research interests are mainly on new synthesis 3D-ordered macroporous and mesoporous catalysts. Cover characterization and application to heterogeneous transformations of a wide range of pure and multi-component metal nanoparticles. Develop colloidal syntheses strategies to synthesize nanoparticles of transition metals. His excellent work has been recognised through several scientific awards, for instance he received the 2013 First Grand Prize *American Dow Sustainability Innovation Challenge Award*, which is highly competitive and aims to support top new investigative scientists from USA and Europe. Other awards he has received include *Outstanding Ph.D. Dissertation* (2014), *Young Scientist Award* Taiwan (2013), and *Certificate of Appreciation Research* by Iran's Ambassador in China (2014).

In 2013, he was a recipient of the youngest ever *Top 10 Student of Tsinghua University* Award. This honor is given to the top 10 students from a pool of 40,000 students across dozens of departments. Among all 4–5th year talented Ph.D. students, second year Ph.D. candidate (Dr. Arandiyan) became the first foreign student to receive this award in 102 years. The award reflects his standing as a significant role model for international students at Tsinghua University, which is considered to be a Top Chinese University. Other honors have followed including (i) "Chinese Government Scholarship-CSC," Ministry of Education of China (2012–2014); (ii) "Comprehensive Postgraduate Scholarship" at THU (2013); (iii) "Outstanding Publication Ph.D. Candidate" (2013); (iv) "Outstanding Qualifying Doctoral Examination" (2012); and (v) "Chinese Scholarship Council" (2011).

Chapter 1
Introduction and Structure of the Thesis

1.1 Research Background

Natural gas burns cleaner than traditional gasoline or diesel properly to its lower carbon substance. Since natural gas used as an automobile gas, it can suggest life series greenhouse gas (GHG) emissions profits over traditional fuels [1]. Depending on automobile type, drive cycle, and engine calibration, the emissions of primary component are related to include the regulated emissions of hydrocarbons (HC), carbon monoxide (CO), oxides of nitrogen (NO_x), and carbon dioxide (CO_2) [2, 3]. In order to include more and more serious emission regulations, the break has been limited between tailpipe emission advantages from NGVs and traditional automobiles with current emission controls. The US Environmental Protection Agency (EPA) is demanding all fuels and automobile types to meet the same onsets for tailpipe emissions of air pollutants. Nevertheless, NGVs persist to supply emission profits. Particularly, when substitute broken traditional automobiles or when taking into consideration life cycle emissions [4]. The overall combustion of CH_4, the primary component of natural gas, is corresponding to the complete reaction as shown in Fig. 1.1.

1.1.1 Significance of Development of Natural Gas Vehicle

Traditional burning of natural gas has some advantages and disadvantages. First, the only products of CH_4 combustion should be carbon dioxide (CO_2) and water (H_2O) according to the complete CH_4 combustion reaction. Since, conventional natural gas combustion frequently use temperatures as high as 2000 K, nitrogen oxides (NO_x) including NO_2 and NO will be formed. These oxides are foundations of acid rain that can bring about serious environmental harm. Natural gas is

H. Arandiyan, *Methane Combustion over Lanthanum-based Perovskite Mixed Oxides*, Springer Theses, DOI 10.1007/978-3-662-46991-0_1

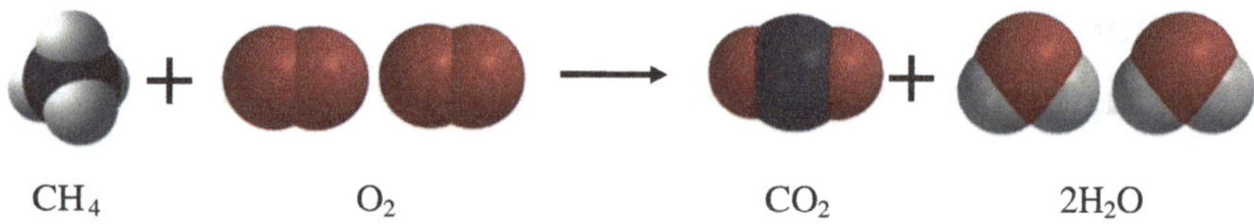

Fig. 1.1 Stoichiometric gas combustion [4]

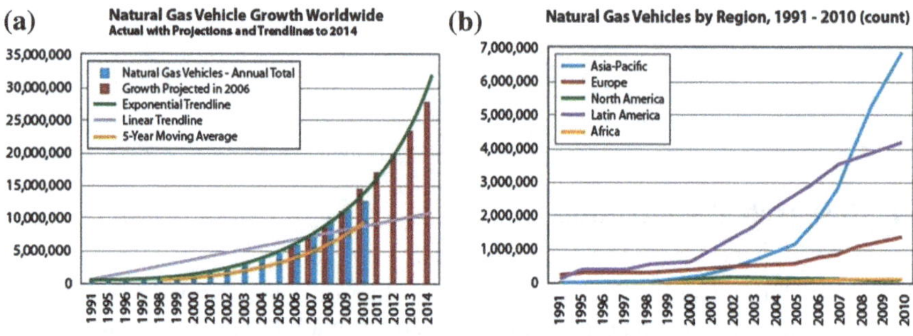

Fig. 1.2 **a** Natural gas vehicle growth worldwide and **b** natural gas vehicles by region [5]

progressively more being used to change gasoline in minor usages, for instance, forklifts and industrial grass apparatus (Fig. 1.2).

1.1.2 Current Situation of the Development of Natural Gas Vehicle

Since natural gas is a low-carbon, clean-burning fuels, a shift to natural gas in these applications might be consequence in considerable reductions of HC, CO, NO_x, and GHGs. In the meantime, natural gas is nonhazardous, so it is not damaging to soil or water. By the end of 2012, there were more than 16.7 million NGVs working in the world. Natural gas is able to be used in most types of automobiles such as motorbikes, vehicles, cars, vans, heavy- and light-duty trucks, buses and mini buses, fork trucks, and locomotives. Oceanic usages are rapidly growing with applications such as tug boats, ferries, barges, and ships using CNG and LNG [5]. Even, NG power-driven aircraft are being trialed.

1.1.3 Control Technology for Emission Pollution of Natural Gas Vehicle

The NGV industries would rise up as producers launched them all over the Asia in an effort to control rising oil request. For the reason that if those populated countries

Table 1.1 European regulation in air quality control from Euro I to Euro VI [5]

Tier	Date	Test cycle	CO	HC	NO_x	PM	Smoke
Euro I	1992, <85 kW	ECE R-49	4.5	1.1	8.0	0.612	
	1992, >85 kW		4.5	1.1	8.0	0.36	
Euro II	October 1996		4.0	1.1	7.0	0.25	
	October 1998		4.0	1.1	7.0	0.15	
Euro III	October 1999	ESC & ELR	1.0	0.25	2.0	0.02	0.15
	October 2000	ESC & ELR	2.1	0.66	5.0	0.10	0.8
Euro IV	October 2005		1.5	0.46	3.5	0.02	0.5
Euro V	October 2008		1.5	0.46	2.0	0.02	0.5
Euro VI	January 2013		1.5	0.13	0.4	0.01	

exposed a desire for petroleum, which would initiate a real problem for the worldwide financial system. In Asia, the average annual growth rate of NGVs has been 42 % over the last 10 years. There are more than a million NGVs in India and 450,000 in China.

Pakistan boasts 2.7 million NGVs, and then, we have Iran, Argentina, and Brazil each with more than 1.5 million apiece. As shown in Table 1.1, hydrocarbon emission strictly controls from Euro I to Euro VI (1.1–0.13). International Association for Natural Gas Vehicles believes the global NGV will improve at least 10 times by 2020 [5].

1.2 Research Progress of Catalytic CH_4 Combustion Catalyst

The oxidation reaction used as oxygen carrier must meet with a number of requirements:

- A high O_2 exchange volume
- Rapid O_2 exchange kinetics at the oxidation reaction temperature
- Should withstand the harsh environment of the partial oxidation reaction
- High-performance activity, or possible to combine with an active phase catalyst
- Reasonably easy to prepare and expenses effective.

According to the listed supplies, perovskite mixed oxides can be suitable type of structures for this goal. Without a doubt, most of the literatures, dealing with CH_4 combustion (as we explain in more detail in the following chapters), concern the use of nonstoichiometric oxides. Generally, one type of nonstoichiometric oxides has been fascinating notice as active structures for cyclic conversion of CH_4 combustion, perovskite mixed oxide [6–10]. A short review of the preview results obtained with this type of structures is given in following sections.

1.2.1 Nonstoichiometry in Perovskite Mixed Oxide (ABO_3)

Perovskite mixed oxides are a type of oxides characterized by the general formula ABO_3. The ideal formation is cubic with the B cation in sixfold coordination surrounded by an octahedron of O_2 atoms and the A cation placed in the large interstices bounded by eight octahedra. The quantity of deformation of the crystal system for each ABO_3 system associates with the Goldschmidt tolerance factor t (Eq. 1.1) [11].

$$(R_A + R_O)/\sqrt{2}(R_B + R_O) = t \quad (1.1)$$

where R_B is the radius of the cation B, R_A the radius of the cation A, and R_O the radius of the anion, usually O_2. Chosen perovskite based on Sr and Co with related tolerance factors (t) based on experimental ionic radii given in preview research groups' jobs [11] is shown in Table 1.2.

While most compounds with perovskite materials are oxides, some carbides, nitrides, halides, and hydrides also form these materials. The Goldschmidt equation is according to the procedure at room temperature. The ideal undistorted cubic structure of $SrTiO_3$ has a tolerance factor of 1. The range of $0.75 < t < 1.02$ corresponds to a distorted perovskite structure which can convert to an ideal cubic structure at high temperatures. About 90 % of all natural metals of the periodic table of elements are known to form a stable perovskite oxide [12]. A sample of the ideal perovskite and distorted orthorhombic structures is shown in Fig. 1.3.

Table 1.2 Perovskite mixed oxides based on $SrCoO_x$ and related tolerance factor based on the Goldschmidt equation [11, 12]

Formula	t	Formula	t	Formula	t	Formula	t
$SrCoO_x$	1.03	$SrFeO_x$	0.98	$LaSrCoO_x$	1.01	$SrFeCoO_x$	1.01
$CaCoO_x$	1.00	$SrMgO_x$	0.95	$CeSrCoO_x$	1.00	$SrMgCoO_x$	0.99
$CeCoO_x$	0.98	$SrMnO_x$	1.01	$PrSrCoO_x$	1.00	$SrMnCoO_x$	1.02
$LaCoO_x$	0.99	$SrMoO_x$	0.98	$NdSrCoO_x$	0.99	$SrMoCoO_x$	1.00
$PbCoO_x$	1.05	$SrNiO_x$	1.03	$CaSrCoO_x$	1.01	$SrNiCoO_x$	1.03
$PrCoO_x$	0.98	$SrPbO_x$	0.92	$PbSrCoO_x$	1.04	$SrPbCoO_x$	0.97
$NdCoO_x$	0.97	$SrPdO_x$	0.96	$SrAlCoO_x$	1.04	$SrPdCoO_x$	0.99
$SrAlO_x$	1.04	$SrSbO_x$	1.00	$SrCrCoO_x$	1.02	$SrSbCoO_x$	1.02
$SrCrO_x$	1.00	$SrSnO_x$	0.96	$SrCuCoO_x$	0.98	$SrSnCoO_x$	0.99
$SrCuO_x$	0.94	$SrZnO_x$	0.94	$SrYCoO_x$	0.95	$SrZnCoO_x$	0.98

Red: Substitution of A-site cation Sr^{2+}; blue: substitution of B-site cation Co^{4+}; dashed: partial substitution of respective cations.

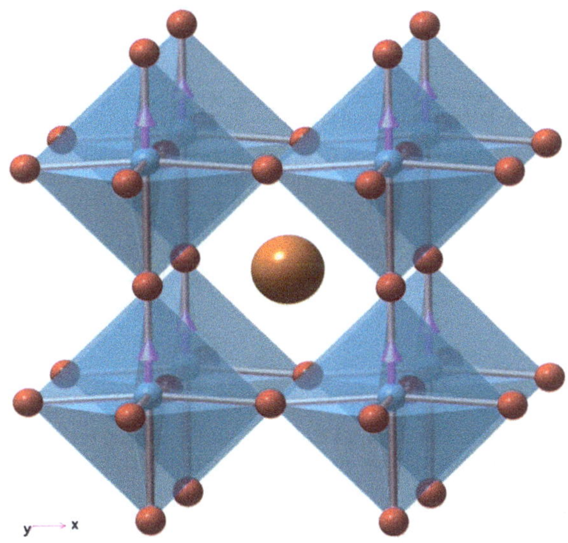

Fig. 1.3 Perovskite structure with chemical formula ABO_3. The *red spheres* are O atoms, the *orange spheres* are B atoms, and the *blue spheres* are the A atoms [13]

1.2.2 Perovskite Metal Oxide Catalysts

The perovskite mixed oxides are very flexible; a number of various compounds can be adapted into the A and B sites. Furthermore, partial substitution of the A and B cations is probable, originating elements like $A_{1-x}A_xB_{1-y}B_yO_3$. Two of the formations that have promising particular concentration for catalytic reactions are those of the so-called LSMO ($La_{1-x}Sr_xMnO_{3-\delta}$) and LSCO ($La_{1-x}Sr_xCoO_{3-\delta}$) categories [14]. The partial substitution of La^{3+} with Sr^{2+} reduces the global charge present on the A site and forces the metal cation in the B site to an oxidation state more than 3+ and may together form O_2 vacancies in the structure [14, 15]. The total O_2 content will in this case turn from the completely stoichiometric value of 3 and suppose a flexible value of $3 - \delta$. The δ denotes the nonstoichiometry value. The O_2 nonstoichiometry is special for each oxide and depends at any moment on the temperature and on the O_2 pressure close to the oxide. Transition metal oxides are promising options because of their thermal stability at high temperatures, flexible composition as well as lower cost for preparing a tailoring potential according to the application. Figure 1.4 compares the catalytic performance of two samples using this indicator; one is $LaFe_{0.57}Co_{0.38}Pd_{0.05}O_3$ (intelligent Pd-perovskite), and the other is conventional Pd/alumina that is Pd-impregnated gamma-Al_2O_3. (The Pd-perovskite catalyst claims its high catalytic performance during aging, while the performance of the Pd/alumina sample decreases by about 10 % [16].)

Perovskite mixed oxides contain in high-temperature solid-state reactions, leading to the damage of pore structures and bring about low BET (<10 m^2/g), unfavorable for development in the catalytic activity of the obtained perovskite

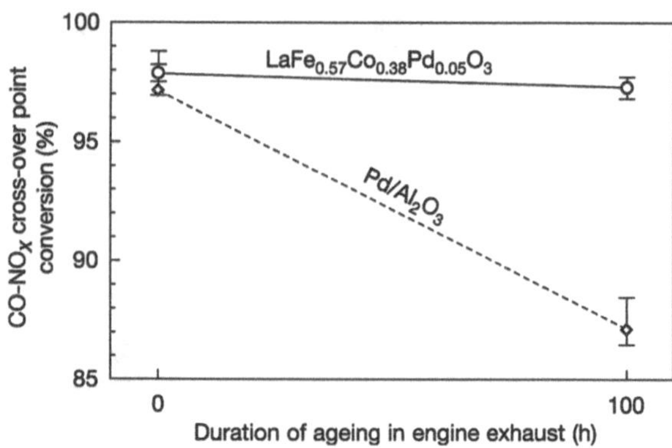

Fig. 1.4 CO–NO_x crossover point conversion for $LaFe_{0.57}Co_{0.38}Pd_{0.05}O_3$ and Pd-impregnated gamma-Al_2O_3 (Pd/alumina catalyst) [16]

structures. For example, the $LaMnO_3$ mixed oxide synthesized by calcination 5 h at 1073 K displayed the lowest BET (9.2 m^2/g) [17], the above values are in good agreement with data reported in the literature. For comparison, some relevant literature data are collected in Table 1.3.

To assess the quality of these perovskite materials, T_{50}, T_{100}, and catalytic performance of other transition metals and the traditional precious metals, including noble-metal-substituted hexa-aluminates studied for methane oxidation in the literature, are shown in Table 1.4. Their performance activity is also comparable to some of the precious metal-substituted oxides.

1.2.3 Partial Substitution Property of Perovskite Structure

A mixture of various transition metals in the B site also raises the performance. Most of the preview jobs in this field concentrated on Mn- and Co-containing oxides in the B site. Because of their promising activity [34, 35], CH_4, the main part of natural gas, is used as a fuel in many combustion procedures, particularly in energy production. Catalytic combustion of CH_4 is beneficial because of lower temperatures desired. As a result, it needs a smaller amount of energy for preheating and produces less or no NO_x. Ferri et al. [36] studied the catalytic oxidation of CH_4 on $La_{1-x}A'_xBO_3$ mixed oxides synthesized by the citrate preparation using Ce, Sr, and Eu precursors in the A site and Fe, Ni, and Co precursors in the B site [36]. They found $La_{0.9}Ce_{0.1}CoO_{3-\delta}$ has high catalytic activity. Other research studies incorporated La-, Ce-, as well as Co-based samples [37–40], La, and La/Sr in the A site and Co, Fe, and Mn in B site [41–44] and partial substitution of La with Ce and Eu [36]. Supported base perovskite oxides (ABO_3, e.g., $LaMnO_3$ [45] and $LaCoO_3$ [46]) are catalytically active in the oxidation of VOCs at high temperatures. It has been confirmed that among the perovskites, the $La_{1-x}Sr_xMO_3$ (M = Co, Mn; x = 0–0.4) structures display the high

Table 1.3 Literature data for the activity in CH_4 combustion over $LaMnO_{3+\lambda}$ perovskite catalysts

References	Preparation method	Calcination (K/h)	BET (m^2/g)	Q^a (ml/min)	GHSV (ml/g h)	K_{823} (μmol/g s bar)
[18]	Coprecipitation	1170/12	8	200	120,000	200[b]
[19]	Solution evaporation	1123/5	4.6	400	45,000	350
[20]	Citrate	1073/5[c]	20	200	40,000	700
[21]	Citrate	1073/5	9.2	75	45,000	424
[22]	Citrate	1073/4	8	72	4320	220
[23]	Spray decomposition	1073/4	8.8	200	12,000	520
[24]	Citrate	1023/2	5	40	12,000	450
[25]	Spray-freezing/ freeze-drying	973/12	13		–	–
[26]	Citrate	973/5	8.8	240	144,000	700
[21]	Citrate	973/5	16	75	45,000	602
[23]	Spray decomposition	973/4	12.2	200	12,000	620
[24]	Citrate	943/2	20	40	12,000	780
LCMO						
[24]	Citrate	1023/2	32	40	12,000	520
LCMO						
[23]	Spray decomposition	973/4	18.7	500	12,000	375

[a]Total flowrate
[b]1 % CH_4/4 % oxygen/helium
[c]Pretreatment 5 h at 823 K, grinding before calcination

catalytic activity [26, 47, 48]. The redox properties of the B-position ions, the nature and concentration of O_2 adspecies, and the existence of lattice defects are usually believed to be responsible for the high catalytic performance of perovskite-type oxide [49]. Agarwal et al. [50] investigated the catalytic oxidation of toluene over $LaMO_3$ (M = Co, Fe, and Cr) synthesized via a decomposition method. They found catalytic performance as follows: $LaCoO_3 > LaFeO_3 > LaCrO_3$. Levasseur and Kaliaguine [51] synthesized a type of $LaBO_3$ (B = Co, Mn, and Fe) samples by adopting a method called the "reactive grinding method." The catalytic activity for methanol oxidation followed the order $LaMnO_3 > LaCoO_3 > LaFeO_3$. Catalytic performance and stability of samples are highly connected to its nature. For instance, surface area, morphology, and crystallinity [52, 53] are influenced by the synthesization route adopted. A big challenge in preparation of ABO_3 with a high BET is to create a porous material. Synthesization of ABO_3 requires a high-temperature solid-state reaction for its metal precursors, which usually brings about the hard preservation of pore structure and finally to a low BET. For that reason, a unique route is strongly recommended to synthesize porous ABO_3 structures with high BET.

Table 1.4 Catalytic activity comparisons in the literatures

Catalyst	Catalyst weight (g) size (mm)	T_{50} (°C)	T_{100} (°C)	A[a] at T_{50}	A[a] at T_{100}	References
$La_{0.9}Ce_{0.1}CoO_3$	0.2 g of powder	450	750	1.48×10^{-4}	2.09×10^{-4}	[27]
$La_{0.1}Ce_{0.4}Co_{1.5}O_3$	0.1–0.25	390	600	2.60×110^{-4}	4.05×10^{-4}	
$La_{0.2}Ce_{0.3}Co_{1.5}O_3$		390	550	2.98×110^{-4}	4.82×10^{-4}	
$La_{0.3}Ce_{0.2}Co_{1.5}O_3$		430	550	3.04×110^{-4}	5.19×10^{-4}	
$La_{0.9}Ce_{0.1}CoO_3$	0.2 g of powder	438	550	4.46×110^{-6}	7.71×10^{-6}	[28]
	0.15–0.25					
$La_{0.7}Ag_{0.3}MnO_3$	2.5 g of powder	~400	~550	1.81×10^{-5}	2.96×10^{-5}	[29]
$La_{0.7}Sr_{0.3}MnO_3$	0.074–0.037	~450	~580	1.69×10^{-5}	2.86×10^{-5}	
$La_{0.7}Ce_{0.3}MnO_3$		~455	~600	1.68×10^{-5}	2.80×10^{-5}	
$La_{0.7}MnO_3$		~460	~600	1.66×10^{-5}	2.79×10^{-5}	
$La_{0.9}Ce_{0.1}CoO_3$	0.2 g of powder	~430	500	8.67×10^{-6}	1.58×10^{-5}	[26]
	0.02–0.04					
1.0 wt% Pd/Al_2O_3	0.1 g of Powder	~367	~600	3.57×10^{-4}	5.24×10^{-4}	[30]
1.0 wt% Pd/ZrO_2	Particle size not mentioned	~317	~527	3.88×10^{-4}	5.72×10^{-4}	
$Sr_{0.8}La_{0.2}MnAl_{11}O_{19}$, commercial	1 g of powder <5 μm	~625	~800	5.43×10^{-5}	9.09×10^{-5}	[31]
1 mol% Pd, LaMnPd	0.6 g of powder	~475	~650	1.13×10^{-5}	1.83×10^{-5}	[32]
1 mol% Pd, LaMnPd	0.1–0.315	~425	~550	1.21×10^{-5}	2.05×10^{-5}	
4 mol% Pd, LaMnPd		~475	~650	1.13×10^{-5}	1.83×10^{-5}	
2 mol% Rh, LaMnRh		~495	~650	1.10×10^{-5}	1.83×10^{-5}	
2 mol% Rh, LaMnRh		~445	~575	1.17×10^{-5}	1.99×10^{-5}	
Co_3O_4	0.1 g of powder	325	420			[33]
0.18 wt% Au/Co_3O_4	0.250–0.177	317	420			
0.21 wt% Pt/Co_3O_4		312	400			
1.92 wt% Pt/Co_3O_4		304	376			
1.58 wt% Pd/Al_2O_3		338	387			
3.70 wt% Pd/Al_2O_3		313	358			

[a]Mass-specific activity [mol (g of catalyst)$^{-1}$ min^{-1}]

Recently, templating routes have been modified for the preparing of porous inorganic materials. For instance, Sadakane et al. [54] synthesized three-dimensionally ordered macroporous (3DOM) $La_{1-x}Sr_xFeO_3$ (x = 0–0.4) with a BET of 24–49 m^2/g using a colloidal crystal polystyrene (PS) microsphere-templating route.

They found that the 3DOM architectures of $LaFeO_3$ catalyst show well combustion of carbon particles. Using the developed polymethyl methacrylate (PMMA) as template, Zhao and coworkers obtained 3DOM $LaCo_xFe_{1-x}O_3$ [55] and 3DOM $LaFeO_3$-supported Au (surface area = 31–32 m^2/g) [56] samples. They found out that super catalytic activity (the temperature required for 50 % soot conversion was 359 °C over 6.3 wt% Au/$LaFeO_3$ and 397 °C over $LaCo_{0.5}Fe_{0.5}O_3$) [57].

1.2.4 Precious Metal Perovskite Catalyst

Gardner et al. [58] investigated the Mn oxide catalysts with low Ag loadings (≤1 %) for low-temperature CO oxidation and found high catalytic performance. Furthermore, Ag-loaded $LaCoO_3$ showed no deactivation in CH_4 combustion during 50 h of onstream reaction at 600 °C [59]. Even better activity improvement in comparison with $LaMnO_3$ was observed in the case of Ag/MnO_x/perovskite composite catalysts [60]. The ABO_3 supports used in most of the literatures, however, are all nonporous. Since porous ABO_3 can offer high dispersion of active components on the surface of catalyst, it would be better to synthesize the Ag/3DOM-ABO_3 architecture that shows large BET and porous structures. To the best of our knowledge, however, there has been no literature on the successful synthesization of Ag NPs supported on three-dimensionally ordered macroporous $La_{0.6}Sr_{0.4}MnO_3$ with mesoporous walls using the PMMA-templating route. And their applications are also in catalyzing the combustion of CH_4. In the past several years, Dai and coworkers have extended attention to the preparation and physicochemical property characterization of 3DOM architecture structures (e.g., Co_3O_4/3DOM $La_{0.6}Sr_{0.4}CoO_3$ (BET = 29–32 m^2/g) [61], yCrOx/3DOM $InVO_4$ (BET = 41.3–52.3 m^2/g) [62], 3DOM $InVO_4$ (BET = 35–52 m^2/g) [63], xAu/3DOM $LaMnO_3$ (BET = 29.8–32.7 m^2/g) [64], Au/3DOM $LaCoO_3$ (BET = 24 −29 m^2/g) [65], and Au/3DOM $La_{0.6}Sr_{0.4}MnO_3$ (BET = 31.1–32.9 m^2/g) [66]) via the surfactant-embed PMMA-templating methods. It observed that some of the 3DOM structure showed excellent catalytic activity in the combustion of CH_4 and toluene [67–69]. In this, we report for the first time the synthesization, characterization, and catalytic CH_4 combustion activities of 3DOM $La_{0.6}Sr_{0.4}MnO_3$ with nanovoid-like or mesoporous skeletons and its supported Ag NPs nanocatalyst (yAg/3DOM $La_{0.6}Sr_{0.4}MnO_3$; y = 0, 1.57, 3.63, and 5.71 wt%). Synthesized by dimethoxytetraethylene glycol (DMOTEG)-mediated gas bubbling reduction method with well-arrayed PMMA microspheres as hard template. It will expect that DMOTEG-assisted reduction strategy not only produced size-controlled Ag NPs, but also stabilized them against conglomeration without the need of additional stabilizers. By using this novel method, Ag NPs could be highly dispersed on 3DOM $La_{0.6}Sr_{0.4}MnO_3$, and the obtained catalysts showed excellent catalytic activity and stability for CH_4 combustion.

1.3 Research Purpose, Idea, and Motivation

1.3.1 Purpose and Significance

Currently, there are no catalyst systems that can be employed in industrial catalytic combustion applications. Catalysts with high-specific activity that are able to low temperatures are not stable in continued operation at natural gas vehicle conditions. On the other hand, stable catalysts that can withstand long-term operation at high temperatures and hydrothermal conditions are not active enough temperatures that would make them economically viable. We have proposed to use nanostructure 3DOM architecture processing to synthesize novel catalyst systems that would allow stable condition and high catalytic performance. Specifically, we have concentrated our efforts on the novel synthesis of pure and modified 3DOM nanostructure perovskite mixed oxide catalysts, such as LSMO ($La_{1-x}Sr_xMnO_{3-\delta}$) and LSCO ($La_{1-x}Sr_xCoO_{3-\delta}$).

Our strategy was to develop a synthesis approach that would generate nonagglomerated nanostructure materials with controlled particle morphology to attain high-BET surface area and sintering resistance and finally to increase the catalytic activity for CH_4 combustion reaction. We would utilize nanocomposite catalyst design to establish synergistic effects between the noble metals, rare earth oxides, or transition metal oxides as well as missed oxide (ABO_3) perovsike on 3DOM structures.

1.3.2 Content and Technical Route

The structure of this thesis consists of three main chapters:

Chapter 3 involves a review of the relevant literature on combustion catalysts. In particular, it details studies of the total oxidation activity of CH_4 combustion performance and their characterization. Developed a novel and facile one-pot hydrothermal approach that develops 1 Dimensional single-crystalline perovskite $La_{0.5}Sr_{0.5}CoO_3$ nanowires without the use of a template, which may be different from those of the nanoparticles. In this chapter, we report the synthesis, characterization, and catalytic performance of nanowire PTOs with dextrose/L-Lysine-assisted hydrothermal route (DHR) for highly performance of methane combustion.

Chapter 4 outlines the preparation and initial characterization of a novel and facile DMOTEG PMMA-templating route. In this chapter, we report the characterization, preparation, and performance activity of CH_4 combustion of 3DOM $La_{0.6}Sr_{0.4}MnO_3$ (3DOM LSMO) with nanovoid-like or mesoporous skeletons by the use of the surfactant (DMOTEG, L-lysine, poly (ethylene glycol), (PEG400), triblock copolymer (Pluronic P-123), and/or ethylene glycol (EG))-assisted PMMA-templating strategy.

Chapter 5 discusses about on catalytic CH_4 combustion activities of 3DOM $La_{0.6}Sr_{0.4}MnO_3$ with nanovoid-like or mesoporous skeletons and its supported Ag NPs nanocatalyst (yAg/3DOM $La_{0.6}Sr_{0.4}MnO_3$; $y = 0$, 1.57, 3.63, and 5.71 wt%) via the DMOTEG-assisted gas bubbling reduction route with well-arrayed PMMA microspheres as hard template. It was found that the DMOTEG-mediated reduction strategy not only produced size-controlled Ag NPs, but also stabilized them against conglomeration without the need of additional stabilizers. By using this novel method, Ag NPs could be highly dispersed on 3DOM $La_{0.6}Sr_{0.4}MnO_3$, and the obtained catalysts showed excellent performance and stability for methane combustion.

References

1. Lee JH, Trimm DL. Catalytic combustion of methane. Fuel Process Technol. 1995;42(2):339–59.
2. Machida M, Eguchi K, Arai H. Catalytic properties of BaMAl11O19-α (M = Cr, Mn, Fe Co, and Ni) for high-temperature catalytic combustion. J Catal. 1989;120(2):377–86.
3. Machida M, Eguchi K, Arai H. Effect of structural modification on the catalytic property of Mn-substituted hexaaluminates. J Catal. 1990;123(2):477–85.
4. Seiyama T. Total oxidation of hydrocarbons on perovskite oxides. Catal Rev. 1992;34(4):281–300.
5. International Association for Natural Gas Vehicles (IANGV).
6. Rida K, Benabbas A, Bouremmad F, Peña MA, Martínez-Arias A. Surface properties and catalytic performance of $La_{1-x}Sr_xCrO_3$ perovskite-type oxides for CO and C_3H_6 combustion. Catal Commun. 2006;7(12):963–8.
7. Wang C-H, Chen C-L, Weng H-S. Surface properties and catalytic performance of $La_{1-x}Sr_xFeO_3$ perovskite-type oxides for methane combustion. Chemosphere. 2004;57(9):1131–8.
8. Alifanti M, Blangenois N, Florea M, Delmon B. Supported Co-based perovskites as catalysts for total oxidation of methane. Appl Catal A. 2005;280(2):255–65.
9. Campagnoli E, Tavares A, Fabbrini L, et al. Effect of preparation method on activity and stability of $LaMnO_3$ and $LaCoO_3$ catalysts for the flameless combustion of methane. Appl Catal B. 2005;55(2):133–9.
10. Yi N, Cao Y, Su Y, Dai W-L, He H-Y, Fan K-N. Nanocrystalline $LaCoO_3$ perovskite particles confined in SBA-15 silica as a new efficient catalyst for hydrocarbon oxidation. J Catal. 2005;230(1):249–53.
11. Arandiyan HR, Parvari M. Preparation of La-Mo-V mixed-oxide systems and their application in the direct synthesis of acetic acid. J Nat Gas Chem. 2008;17(3):213–24.
12. Khalesi A, Arandiyan HR, Parvari M. Production of Syngas by CO_2 Reforming on $MxLa_{1-x}Ni_{0.3}Al_{0.7}O_{3-d}$ (M = Li, Na, K) Catalysts. Ind Eng Chem Res. 2008;47(16):5892–8.
13. Arandiyan HR, Parvari M. Studies on mixed metal oxides solid solutions as heterogeneous catalysts. Braz J Chem Eng. 2009;26:63–74.
14. Deng J, Zhang Y, Dai H, Zhang L, He H, Au CT. Effect of hydrothermal treatment temperature on the catalytic performance of single-crystalline $La_{0.5}Sr_{0.5}MnO_{3-\delta}$ microcubes for the combustion of toluene. Catal Today. 2008;139(1):82–7.
15. Pecchi G, Jiliberto MG, Delgado EJ, Cadús LE, Fierro JLG. Effect of B-site cation on the catalytic activity of $La_{1-x}Ca_xBO_3$ (B = Fe, Ni) perovskite-type oxides for toluene combustion. J Chem Technol Biotechnol. 2011;86(8):1067–73.

16. Nishihata Y, Mizuki J, Akao T, et al. Self-regeneration of a Pd-perovskite catalyst for automotive emissions control. Nature. 2002;418(6894):164–7.
17. Najjar H, Lamonier J-F, Mentré O, Giraudon J-M, Batis H. Optimization of the combustion synthesis towards efficient $LaMnO_{3+y}$ catalysts in methane oxidation. Appl Catal B. 2011;106 (1):149–59.
18. Marti PE, Baiker A. Influence of the A-site cation in $AMnO_{3+x}$ and $AFeO_{3+x}$ (A = La, Pr, Nd and Gd) perovskite-type oxides on the catalytic activity for methane combustion. Catal Lett. 1994;26(1):71–84.
19. Arai H, Yamada T, Eguchi K, Seiyama T. Catalytic combustion of methane over various perovskite-type oxides. Appl Catal. 1986;26(1):265–76.
20. Ciambelli P, Cimino S, De Rossi S, et al. $AMnO_3$ (A = La, Nd, Sm) and $Sm_{1-x}Sr_xMnO_3$ perovskites as combustion catalysts: structural, redox and catalytic properties. Appl Catal B. 2000;24(3):243–53.
21. Alifanti M, Kirchnerova J, Delmon B. Effect of substitution by cerium on the activity of $LaMnO_3$ perovskite in methane combustion. Appl Catal A Gen. 2003;245(2):231–44.
22. Saracco G, Geobaldo F, Baldi G. Methane combustion on Mg-doped $LaMnO_3$ perovskite catalysts. Appl Catal B. 1999;20(4):277–88.
23. Song K-S, Xing Cui H, Kim SD, Kang S-K, Catalytic combustion of CH_4 and CO on $La_{1-x}M_xMnO_3$ perovskites. Catal Today. 1999;47(1):155–60.
24. Marchetti L, Forni L. Catalytic combustion of methane over perovskites. Appl Catal B. 1998;15(3):179–87.
25. Ng Lee Y, El-Fadli Z, Sapiña F, Martinez-Tamayo E, Cortés Corberán V. Synthesis and surface characterization of nanometric $La_{1-x}K_xMnO_{3+\delta}$ particles. Catal Today. 1999; 52 (1):45–52.
26. de Araujo GC, Lima S, Rangel MdC, Parola VL, Peña MA, García Fierro JL. Characterization of precursors and reactivity of $LaNi_{1-x}Co_xO_3$ for the partial oxidation of methane. Catal Today. 2005; 107(1):906–912.
27. Machin NE, Karakaya C, Celepci A. Catalytic Combustion of Methane on La-, Ce-, and Co-Based Mixed Oxides. Energy Fuels. 2008;22(4):2166–71.
28. Cimino S, Lisi L, Pirone R, Russo G, Turco M. Methane combustion on perovskites-based structured catalysts. Catal Today. 2000;59(1):19–31.
29. Tanaka H, Misono M. Advances in designing perovskite catalysts. Curr Opin Solid State Mater Sci. 2001;5(5):381–7.
30. Yang S, Maroto-Valiente A, Benito-Gonzalez M, Rodriguez-Ramos I, Guerrero-Ruiz A. Methane combustion over supported palladium catalysts: I. Reactivity and active phase. Appl Catal B. 2000;28(3):223–33.
31. Kirchnerova J, Klvana D. Design criteria for high-temperature combustion catalysts. Catal Lett. 2000;67(2):175–81.
32. Giebeler L, Kießling D, Wendt G. $LaMnO_3$ perovskite supported noble metal catalysts for the total oxidation of methane. Chem Eng Technol. 2007;30(7):889–94.
33. Miao S, Deng Y. Au–Pt/Co_3O_4 catalyst for methane combustion. Appl Catal B. 2001;31(3): L1–4.
34. Wang Y, Ren J, Wang Y, et al. Nanocasted Synthesis of mesoporous $LaCoO_3$ perovskite with extremely high surface area and excellent activity in methane combustion. J Phys Chem C. 2008;112(39):15293–8.
35. Liu Y, Zheng H, Liu J, Zhang T. Preparation of high surface area $La_{1-x}A_xMnO_3$ (A = Ba, Sr or Ca) ultra-fine particles used for CH_4 oxidation. Chem Eng J. 2002;89(1):213–21.
36. Borovskikh L, Mazo G, Kemnitz E. Reactivity of oxygen of complex cobaltates $La_{1-x}Sr_xCoO_{3-\delta}$ and $LaSrCoO_4$. Solid State Sci. 2003;5(3):409–17.
37. Saracco G, Scibilia G, Iannibello A, Baldi G. Methane combustion on Mg-doped $LaCrO_3$ perovskite catalysts. Appl Catal B. 1996;8(2):229–44.
38. Belessi VC, Ladavos AK, Pomonis PJ. Methane combustion on La-Sr-Ce-Fe-O mixed oxides: bifunctional synergistic action of $SrFeO_{3-x}$ and CeO_x phases. Appl Catal B. 2001;31 (3):183–94.

39. Kaliaguine S, Van Neste A, Szabo V, Gallot JE, Bassir M, Muzychuk R. Perovskite-type oxides synthesized by reactive grinding: Part I. Preparation and characterization. Appl Catal A. 2001;209(1):345–58.
40. Spinicci R, Tofanari A, Delmastro A, Mazza D, Ronchetti S. Catalytic properties of stoichiometric and non-stoichiometric $LaFeO_3$ perovskite for total oxidation of methane. Mater Chem Phys. 2002;76(1):20–5.
41. Alifanti M, Kirchnerova J, Delmon B, Klvana D. Methane and propane combustion over lanthanum transition-metal perovskites: role of oxygen mobility. Appl Catal A. 2004;262 (2):167–76.
42. Machocki A, Ioannides T, Stasinska B, et al. Manganese–lanthanum oxides modified with silver for the catalytic combustion of methane. J Catal. 2004;227(2):282–96.
43. Campagnoli E, Tavares A, Fabbrini L, et al. Effect of preparation method on activity and stability of $LaMnO_3$ and $LaCoO_3$ catalysts for the flameless combustion of methane. Appl Catal B. 2005;55(2):133–9.
44. Royer S, Alamdari H, Duprez D, Kaliaguine S. Oxygen storage capacity of $La_{1-x}BO_3$ perovskites relation with catalytic activity in the CH_4 oxidation reaction. Appl Catal B. 2005;58(3):273–88.
45. Wyrwalski F, Lamonier JF, Siffert S, Aboukaïs A. Additional effects of cobalt precursor and zirconia support modifications for the design of efficient VOC oxidation catalysts. Appl Catal B. 2007;70(1):393–9.
46. Irusta S, Pina MP, Menéndez M, Santamaría J. Catalytic combustion of volatile organic compounds over La-based perovskites. J Catal. 1998;179(2):400–12.
47. Deng M-J, Ho P-J, Song C-Z, et al. Fabrication of Mn/Mn oxide core-shell electrodes with three-dimensionally ordered macroporous structures for high-capacitance supercapacitors. Energy Environ Sci. 2013;6(7):2178–85.
48. Niu J, Deng J, Liu W, et al. Nanosized perovskite-type oxides $La_{1-x}Sr_xMO_{3-\delta}$ for the catalytic removal of ethylacetate. Catal Today. 2007;126(3):420–9.
49. O'Connell M, Norman AK, Hüttermann CF, Morris MA. Catalytic oxidation over lanthanum-transition metal perovskite materials. Catal Today. 1999;47(1):123–32.
50. Agarwal DD, Goswami HS. Toluene oxidation on $LaCoO_3$, $LaFeO_3$ and $LaCrO_3$ perovskite catalysts. A comparative study. React Kinet Catal Lett. 1994;53(2):441–9.
51. Levasseur B, Kaliaguine S. Methanol oxidation on $LaBO_3$ (B = Co, Mn, Fe) perovskite-type catalysts prepared by reactive grinding. Appl Catal A. 2008;343(1–2):29–38.
52. Campagnoli E, Tavares A, Fabbrini L, et al. Effect of preparation method on activity and stability of $LaMnO_3$ and $LaCoO_3$ catalysts for the flameless combustion of methane. Appl Catal B. 2005;55(2):133–9.
53. Yamazoe N, Teraoka Y. Oxidation catalysis of perovskites relationships to bulk structure and composition (valency, defect, etc.). Catal Today. 1990;8(2):175–99.
54. Sadakane M, Asanuma T, Kubo J, Ueda W. Facile procedure to prepare three-dimensionally ordered macroporous (3DOM) perovskite-type mixed metal oxides by colloidal crystal templating method. Chem Mater. 2005;17(13):3546–51.
55. Xu J, Liu J, Zhao Z, et al. Three-dimensionally ordered macroporous $LaCo_xFe_{1-x}O_3$ perovskite-type complex oxide catalysts for diesel soot combustion. Catal Today. 2010;153 (3):136–42.
56. Wei Y, Liu J, Zhao Z, et al. Highly active catalysts of gold nanoparticles supported on three-dimensionally ordered macroporous $LaFeO_3$ for soot oxidation. Angew Chem. 2011;123 (10):2374–7.
57. Wei Y, Liu J, Zhao Z, et al. Three-dimensionally ordered macroporous $Ce_{0.8}Zr_{0.2}O_2$-supported gold nanoparticles: synthesis with controllable size and super-catalytic performance for soot oxidation. Energy Environ Sci. 2011;4(8):2959–70.
58. Gardner SD, Hoflund GB, Schryer DR, Schryer J, Upchurch BT, Kielin EJ. Catalytic behavior of noble metal/reducible oxide materials for low-temperature CO oxidation. 1. Comparison of catalyst performance. Langmuir. 1991;7(10):2135–9.

59. Choudhary VR, Uphade BS, Pataskar SG, Thite GA. Low-temperature total oxidation of methane over Ag-doped $LaMO_3$ perovskite oxides. Chem Commun. 1996;9:1021–2.
60. Song KS, Kang SK, Kim SD. Preparation and characterization of Ag/MnO_x/perovskite catalysts for CO oxidation. Catal Lett. 1997;49(1–2):65–8.
61. Li X, Dai H, Deng J, et al. In situ PMMA-templating preparation and excellent catalytic performance of Co_3O_4/3DOM $La_{0.6}Sr_{0.4}CoO_3$ for toluene combustion. Appl Catal A. 2013;458(1):11–20.
62. Wang Y, Dai H, Deng J, et al. 3DOM $InVO_4$-supported chromia with good performance for the visible-light-driven photodegradation of rhodamine B. Solid State Sci. 2013;24(1):62–70.
63. Wang Y, Dai H, Deng J, et al. Three-dimensionally ordered macroporous $InVO_4$: fabrication and excellent visible-light-driven photocatalytic performance for methylene blue degradation. Chem Eng J. 2013;226(1):87–94.
64. Liu Y, Dai H, Deng J, et al. PMMA-templating generation and high catalytic performance of chain-like ordered macroporous $LaMnO_3$ supported gold nanocatalysts for the oxidation of carbon monoxide and toluene. Appl Catal B. 2013;140(1):317–26.
65. Li X, Dai H, Deng J, et al. Au/3DOM $LaCoO_3$: high-performance catalysts for the oxidation of carbon monoxide and toluene. Chem Eng J. 2013;228(1):965–75.
66. Liu Y, Dai H, Deng J, et al. Au/3DOM $La_{0.6}Sr_{0.4}MnO_3$: highly active nanocatalysts for the oxidation of carbon monoxide and toluene. J Catal. 2013;305(1):146–53.
67. Liu Y, Dai H, Du Y, et al. Controlled preparation and high catalytic performance of three-dimensionally ordered macroporous $LaMnO_3$ with nanovoid skeletons for the combustion of toluene. J Catal. 2012;287(1):149–60.
68. Yuan J, Dai H, Zhang L, et al. PMMA-templating preparation and catalytic properties of high-surface-area three-dimensional macroporous La_2CuO_4 for methane combustion. Catal Today. 2011;175(1):209–15.
69. Arandiyan H, Dai H, Deng J, et al. Three-dimensionally ordered macroporous $La_{0.6}Sr_{0.4}MnO_3$ with high surface areas: active catalysts for the combustion of methane. J Catal. 2013;307 (1):327–39.

Chapter 2
Experimental Materials and Methods

2.1 Preparation of Catalyst

2.1.1 Main Chemical Reagent

The experimental work in this PhD study is based on the use of several characterization techniques. These techniques are briefly described in the following chapters. The focus is set on the reasons for selecting the various techniques for this study, accompanied by a general description of each technique. A detailed description of the different measurement parameters and equipments can be found in the experimental section of the articles that compose this PhD work. The chemical reagent materials used in this study are outlined in Table 2.1. All of the chemicals (A.R. in purity) were purchased from Tianjin Guangfu Fine Chemical Research Institute and used without further purification.

2.1.2 Main Equipments

The main equipments and gases for reaction experiments used in this research are summarized in Table 2.2.

2.1.3 Preparation of 1D Single-Crystalline LSCO Nanowires

The LSCO catalysts were prepared by means of the DHR strategy with metal nitrates as the precursor. In a typical preparation process, 0.01 mol of lanthanum (III) nitrate hexahydrate, strontium nitrate, and cobalt nitrate was first dissolved in 25.0 ml of deionized (DI) water under magnetic stirring. After being well mixed,

H. Arandiyan, *Methane Combustion over Lanthanum-based Perovskite Mixed Oxides*, Springer Theses, DOI 10.1007/978-3-662-46991-0_2

Table 2.1 Chemicals and gases used in the experiments

Item	Chemical reagent	Molecular formula
1	Deionized (DI) water	H_2O
2	Lanthanum(III) nitrate hexahydrate	$La(NO_3)_3 \cdot 6H_2O$
3	Strontium nitrate	$Sr(NO_3)_2$
4	Manganese(II) nitrate	$Mn(NO_3)_2$ (50 wt% aqueous solution)
5	Citric acid	$C_6H_8O_7$
6	Dextrose	$C_6H_{12}O_6$
7	L-Lysine	$HO_2CCH(NH_2)(CH_2)_4NH_2$
8	Nitric acid	HNO_3 (38 wt%) solution
9	Potassium peroxydisulfate	$K_2S_2O_8$
10	Methyl methacrylate	$CH_2{=}C(CH_3)COOCH_3$
11	Pluronic P-123	$HO(CH_2CH_2O)_{20}(CH_2CH(CH_3)O)_{70}(CH_2CH_2O)_{20}H$
12	L-Lysine	$HO_2CCH(NH_2)(CH_2)_4NH_2$
13	PEG400	$C_{2n}H_{4n+2}O_{n+1}$, $n = 8.2–9.1$
14	Methanol	CH_3OH
15	Ethylene glycol	$C_2H_6O_2$
16	DMOTEG	Dimethoxytetraethylene glycol
17	Aqueous solution	$HAgCl_4$ aqueous solution
18	Sodium borohydride	$NaBH_4$

Table 2.2 Apparatus used in the experiments

Item	Equipments
1	Magnetic stirring
2	pH value
3	Teflon-lined stainless steel autoclave
4	Buchner funnel
5	Furnace
6	Oven
7	GC-Agilent 7890A
8	Porapack-Q/molecular sieve 5A
9	RT-QPlot divinylbenzene PLOT
10	Fixed-bed quartz tubular microreactor (6 mm)
11	CH_4/He (4.95 %)
12	N_2 (99.999 %)
13	O_2 (99.999 %)
14	SO_2/He (5000 ppm)
15	He (99.999 %)
16	Air

Table 2.3 Preparation parameters of the as-prepared $La_{0.5}Sr_{0.5}CoO_3$ catalysts

Catalyst code	Preparation method	Solution	Calcination condition
LSCO-0	Hydrothermal 170 °C, 30 h	0.1 mol/l Dex + 0.5 g L-Lysine	300 °C, N_2 (20 ml/min), 2 h → 800 °C, air (50 ml/min), 4 h
LSCO-1	Hydrothermal 170 °C, 30 h	0.3 mol/l Dex + 0.5 g L-Lysine	300 °C, N_2 (20 ml/min), 2 h → 800 °C, air (50 ml/min), 4 h
LSCO-2	Hydrothermal 170 °C, 30 h	0.3 mol/l Dex + 1.0 g L-Lysine	300 °C, N_2 (20 ml/min), 2 h → 800 °C, air (50 ml/min), 4 h
LSCO-3	Hydrothermal 170 °C, 30 h	0.3 mol/l Dex + 3.0 g L-Lysine	300 °C, N_2 (20 ml/min), 2 h → 800 °C, air (50 ml/min), 4 h
LSCO-4	Sol–gel	Deionized (DI) water	800 °C, air (50 ml/min), 4 h

specific amounts of $C_6H_8O_7$ (total metal/citric acid molar ratio = 1/1), dextrose, and L-Lysine were in order added to the above mixed solution. Then, the nitric acid (38 wt%) solution was added dropwise to adjust the pH value (ca. 5). After being diluted to a total volume of 40.0 ml, the mixture was then transferred into a 50-ml Teflon autoclave.

Then start hydrothermal treatment which placed in an oven at 170 °C for 1800 min. As shown in Table 2.3, the catalysts fabricated at different dextrose/L-Lysine volumetric ratio 0.1/0.5, 0.3/0.5, 0.3/1.0, and 0.3/3.0 were denoted as LSCO-0, LSCO-1, LSCO-2, and LSCO-3, respectively. When the autoclave was cooled RT, the obtained precursor was first dried at 120 °C overnight and then well ground. As a final point, the obtained black powders were calcined in a N_2 flow of 20 ml/min at a rate of 1 °C/min from RT to 300 °C. Maintained at this temperature for 120 min, and further to 800 °C and kept at this temperature for 240 min in an airflow of 50 ml/min. For comparison, we also prepared a polycrystalline $La_{0.5}Sr_{0.5}CoO_3$ catalyst (calcined at 800 °C for 240 min) by means of sol–gel method [1] which was denoted as LSCO-4 (as shown in Scheme 2.1).

2.1.4 *Synthesis of Monodisperse PMMA Microspheres*

Monodisperse (PMMA) microspheres with an average diameter of ca. 291 nm (Figs. 2.1 and 2.2) were synthesized adopting the procedures described in the literature [2–4]. 0.40 g (3.00 mmol) of potassium peroxydisulfate ($K_2S_2O_8$) and 1500 ml of deionized water were mixed under stirring at 400 rpm, heated at 70 °C, and degassed with flowing N_2 in a separable four-necked 2000-ml round-bottom flask. After equilibrating to 70 °C, 115 ml of methyl methacrylate was poured into the flask, and the resulting suspension was stirred at 70 °C for 60 min. The PMMA colloidal crystal template was prepared with 120 ml of the colloidal suspension (ca. 10.0 g) in a 25-ml centrifugation tube for 75 min (Fig. 2.1).

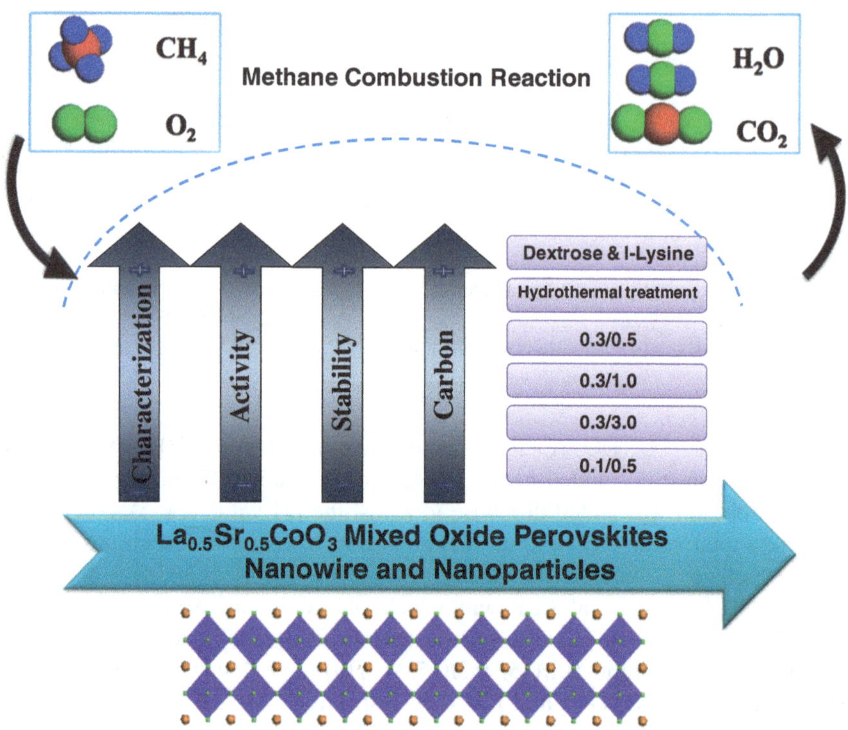

Scheme 2.1 Schematic illustration of 1D single-crystalline LSCO nanowire application

When the DI water was evaporated in water bath at 80 °C, the obtained wet dense material was first dried at RT for two days and then well ground. During the polymerization, several factors (such as monomer, temperature, initiators, and the stirring rate) could influence the PMMA microspheres' sizes.

2.1.5 Preparation of Three-Dimensionally Ordered Macroporous LSMO

A series of 3DOM LSMO catalysts with mesoporous walls were synthesized by the DMOTEG-assisted PMMA-templating strategy with nitrites of lanthanum(III) nitrate hexahydrate, strontium nitrate, and cobalt nitrate as metal precursors as well as present or absence of L-lysine, PEG400, Pluronic P-123, EG, methanol (MeOH), and DMOTEG [5–10]. The typical synthesis procedures are as follows:

(i) 1.00, 3.00, or 5.00 ml of DMOTEG, 5.00 ml of PEG400, 7.79 g of lanthanum(III) nitrate hexahydrate, 2.53 g of strontium nitrate, and 6.97 ml of manganese nitrate (50 wt% aqueous solution) were dissolved in 9.00 ml of DI

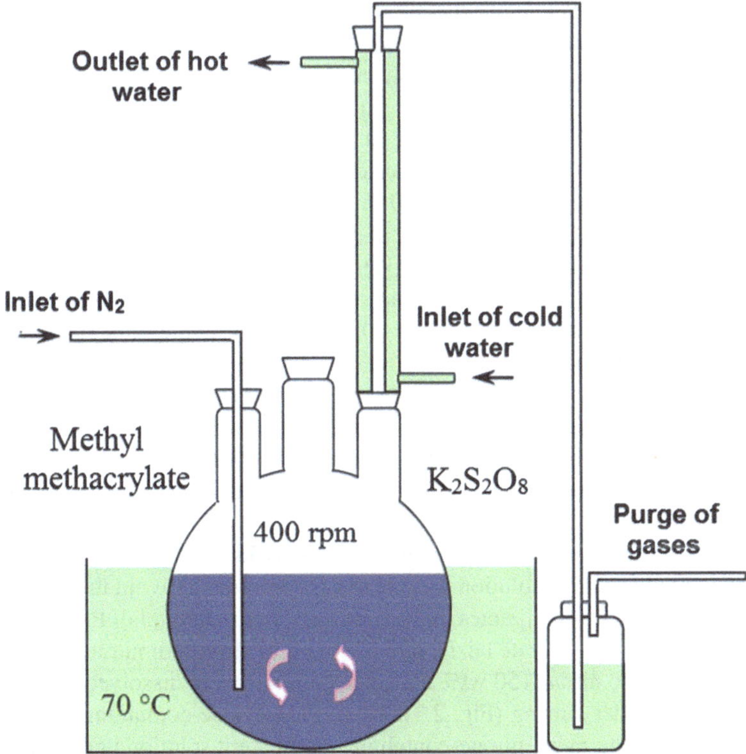

Fig. 2.1 The schematic equipment used for the synthesis of monodispersive PMMA microbeads

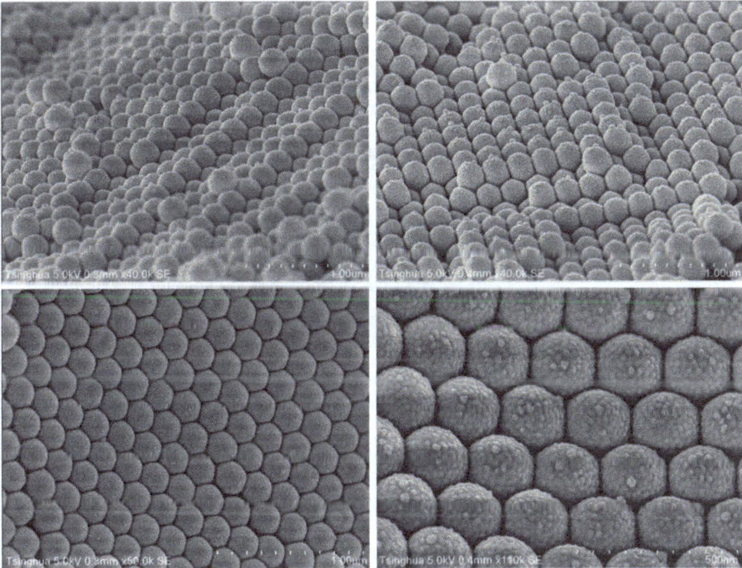

Fig. 2.2 As-prepared monodisperse (PMMA) microspheres

at room temperature under stirring for 240 min to obtain a clear solution. After obtaining a homogeneous precursor solution, ca. 2.00 g of highly ordered PMMA colloidal crystal microspheres was added and flooded with the above mixed solution. Then, excessive liquid was filtered via a Buchner funnel vacuum (0.07 MPa), after dried at room temperature for overnight. The 3DOM materials were reached by calcination according to the two procedures: (a) burned in a nitrogen gas of 50 ml/min at a ramp of 1 °C/min from room temperature to 300 °C. Then held at 300 °C for 180 min; and (b) after being chilled to 30 °C in the same atmosphere and purged in an airflow of 50 ml/min. Then, solid was burned at a ramp of 1 °C/min from room temperature to 300 °C and then held at 300 °C for 60 min. Finally, at the same ramp from 300 to 800 °C and maintained at 800 °C for 240 min (Scheme 2.2).
The obtained samples were, respectively, denoted as LSMO-DP1, LSMO-DP3, and LSMO-DP5. In order to be able to examine the study on calcination temperature on the pore structure, the LSMO-DP3 sample calcined at 850 and 900 °C, thus reaching the LSMO-DP3-850 and LSMO-DP3-900 samples, respectively.

(ii) 1.00 g of L-lysine was dissolved in a nitric acid (38 wt%) solution (5 mol/l), and the pH value of this solution was set to ca. 6 in order to avoid the formation of metal hydroxide precipitates in the following steps. 3.00 ml of PEG400, 0.79 g of lanthanum(III) nitrate hexahydrate, 2.53 g of strontium nitrate, and 6.97 ml of manganese nitrate (50 wt% aqueous solution) were dissolved in 9.00 ml of DI water under stirring (Fig. 2.3). Then, the L-lysine-containing solution was mixed with the metal nitrate-containing transparent solution under stirring for 60 min to get a uniform precursor solution. After dissolution, 2.00 g of highly

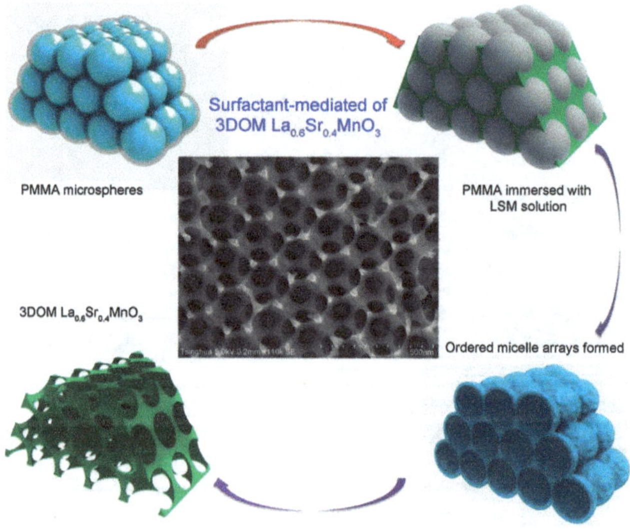

Scheme 2.2 Schematic illustration of the preparation of 3DOM catalyst

Fig. 2.3 Illustrate the procedure of preparation 3DOM catalyst

ordered PMMA colloidal crystal microspheres was added. The subsequent procedures were the same as those described in (i). The obtained catalysts were denoted as LSMO-LP1, in which the "LP" represents the use of "L-lysine" and "PEG400" as surfactant.

(iii) 5.00 ml of EG, 3.00 ml of MeOH, 1.20 g of Pluronic P-123 ($MW_{aver.}$ = 5800 g/mol), 7.79 g of lanthanum(III) nitrate hexahydrate, 2.53 g of strontium nitrate, and 6.97 ml of manganese nitrate (50 wt% aqueous solution) were dissolved in 9.00 ml of deionized water under stirring. After dissolution, 2.00 g of PMMA colloidal crystal microspheres was added. The subsequent processes were the same as those described in (ii). The obtained samples were denoted as LSMO-PE1 (Table 2.4).

2.1.6 Preparation of ywt% Ag/3DOM LSMO Series

The $La_{0.4}Sr_{0.6}MnO_3$-supported silver (*y*Ag/3DOM $La_{0.4}Sr_{0.6}MnO_3$) catalysts were prepared via a DMOTEG-assisted reduction method. During the homogeneous catalysis of methane combustion by Ag NPs supported on 3DOM $La_{0.4}Sr_{0.6}MnO_3$, we observed that DMOTEG was used as a multifunctional reagent for the formation of Ag NPs. It was found that the DMOTEG-mediated route not only produced size-controlled Ag NPs, but also stabilized them against conglomeration without the need

Table 2.4 Preparation parameters of the 3DOM LSMO and 1D LSMO samples

Catalyst code	Surfactant	Soaking time (h)	Calcination condition
1D LSMO	–	–	850 °C 4 h (in air)
LSMO-DP1	1.0 ml DMOTEG + 5.0 ml PEG400 + 9.0 ml H_2O	4	300 °C 3 h (N_2) → 300 °C 1 h (air) → 800 °C 4 h (air)
LSMO-DP3	3.0 ml DMOTEG + 5.0 ml PEG400 + 9.0 ml H_2O	4	300 °C 3 h (N_2) → 300 °C 1 h (air) → 800 °C 4 h (air)
LSMO-DP5	5.0 ml DMOTEG + 5.0 ml PEG400 + 9.0 ml H_2O	4	300 °C 3 h (N_2) → 300 °C 1 h (air) → 800 °C 4 h (air)
LSMO-LP1	1.0 g L-lysine + 1.4 ml HNO_3 + 3.0 ml PEG400 + 9.0 ml H_2O	3	300 °C 3 h (N_2) → 300 °C 1 h (air) → 800 °C 4 h (air)
LSMO-PE1	1.2 g P123 + 5.0 ml EG + 3.0 ml MeOH + 9.0 ml H_2O	3	300 °C 3 h (in N_2) → 300 °C 1 h (air) → 800 °C 4 h (air)
LSMO-DP3-850	1.0 ml DMOTEG + 5.0 ml PEG400 + 9.0 ml H_2O	4	300 °C 3 h (in N_2) → 300 °C 1 h (air) → 850 °C 4 h (air)
LSMO-DP3-900	1.0 ml DMOTEG + 5.0 ml PEG400 + 9.0 ml H_2O	4	300 °C 3 h (in N_2) → 300 °C 1 h (air) → 900 °C 4 h (air)

for additional stabilizers. Thus, we present a facile DMOTEG-mediated synthesis of Ag NPs in the absence of any extra reducing or capping agent. The typical preparation procedure is as follows: A desired amount of PEG (MW = 400 g/mol) and DMOTEG (MW = 222.28 g/mol) was added to a 100 mg/L $HAgCl_4$ aqueous solution (Ag/DMOTEG mass ratio = 1.5:1) at RT under vigorous bubbling for 10 min. After rapid injection of an aqueous solution of 0.1 mol/l $NaBH_4$ (Ag/$NaBH_4$ molar ratio = 1:5), one could obtain a dark brown solution (so-called silver sol). A desired amount of the $La_{0.4}Sr_{0.6}MnO_3$-supported silver was then added to the silver sol (theoretical Ag loading = 2, 4, or 6 wt%), and the obtained suspension was subjected to sonication (60 kHz) for 30 s. A gas-bubble-assisted stirring operation with three bubble outlets in solution was used to further agitate the system, and the suspension was vigorously bubbled with N_2 for 8 h. The solid was collected by filtration, followed by washing with 2.0 L of deionized water. After obtaining a uniform precursor solution, ca. 3.00 g of highly ordered PMMA colloidal crystal microspheres was added and soaked with the above mixed solution. After the PMMA microspheres were thoroughly wet, the excessive liquid was filtered via a Buchner funnel connected to vacuum (0.07 MPa). After being dried at room temperature for 24 h, the 3DOM materials were thermally treated in a tubular furnace according to the two steps: (a) heating in a N_2 flow of 50 ml/min at a ramp of 1 °C/min from RT to 300 °C and keeping it at this temperature for 3 h; and (b) after being cooled to 30 °C in the same atmosphere and purged in an airflow of 50 ml/min, the solid was heated at a ramp of 1 °C/min from RT to 300 °C and kept at this temperature for 1 h, finally same ramp 300–800 °C and maintained at 800 °C for 4 h (Scheme 2.3).

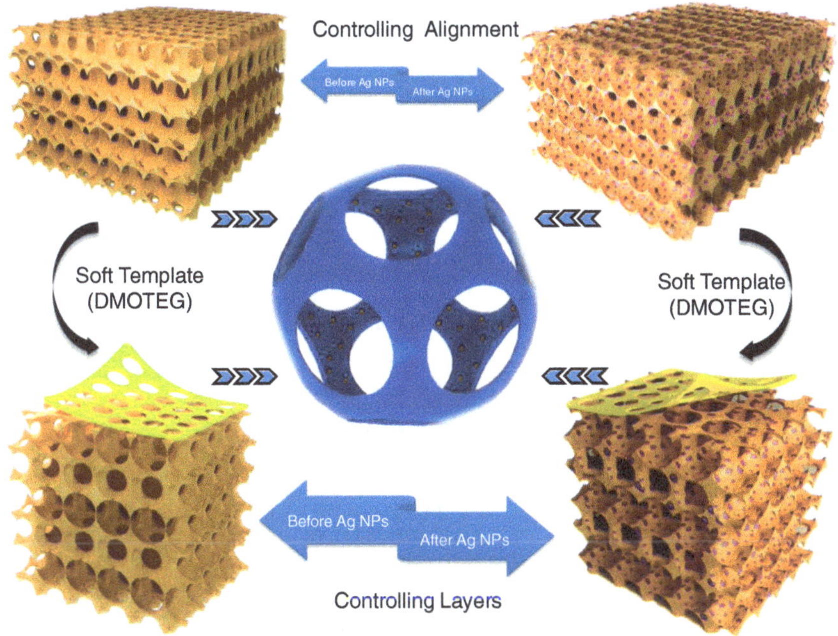

Scheme 2.3 Schematic illustration of loading Ag NPs on 3DOM $La_{0.6}Sr_{0.4}MnO_3$ catalysts

The obtained samples were, respectively, denoted as *y*Ag/3DOM $La_{0.6}Sr_{0.4}MnO_3$. The results of inductively coupled plasma atomic emission spectroscopy (ICP-AES) investigation reveal that the real Ag loading (*y*) was 1.58, 3.63, and 5.71 wt%, respectively (Table 5.1). For comparison purposes, we also prepared 1D $La_{0.6}Sr_{0.4}MnO_3$ sample. The 1D $La_{0.6}Sr_{0.4}MnO_3$ sample was prepared using the citric acid-complexing method described elsewhere [11–14]. The obtained solid precursor was first calcined in air at a ramp of 1 °C/min in a muffle furnace from RT to 300 °C and kept at this temperature for 1 h and then from 300 to 800 °C at the same ramp and maintained at 800 °C for 4 h (Table 2.5).

2.2 Catalytic Activity Measurement

2.2.1 Gas Flow Measurement

Catalytic performance assessment was carried out in a fixed-bed quartz tubular microreactor (i.d. = 6.0 mm), as shown in Fig. 2.4. To minimize the effect of hot spots, the sample (20 mg, 40–60 mesh) was diluted with 0.25 g quartz sands (40–60 mesh). The volumetric composition of the reactant mixture was 2 % CH_4 + 20 % O_2 + 78 % N_2 (balance), and the total flow was 41.6 ml/min, thus giving a gas hourly space velocity (GHSV) of ca. 30,000 ml/(g h).

Table 2.5 Preparation conditions of the 1D $La_{0.6}Sr_{0.4}MnO_3$, 3DOM $La_{0.6}Sr_{0.4}MnO_3$, and *y*Ag/3DOM $La_{0.6}Sr_{0.4}MnO_3$ samples

Catalyst	Method	Hard template/soft template	Calcination condition
1D $La_{0.6}Sr_{0.4}MnO_3$	Citric acid	–/–	850 °C, air (50 ml/min), 4 h
3DOM $La_{0.6}Sr_{0.4}MnO_3$	PMMA templating	PMMA/ (DMOTEG-PEG400)	300 °C, N_2 (50 ml/min), 3 h → 300 °C, air (50 ml/min), 1 h → 800 °C, air (50 ml/min), 4 h
1.58 wt% Ag/3DOM $La_{0.6}Sr_{0.4}MnO_3$	In situ PMMA templating	PMMA/ (DMOTEG-PEG400)	300 °C, N_2 (50 ml/min), 3 h → 300 °C, air (50 ml/min), 1 h → 800 °C, air (50 ml/min), 4 h
3.63 wt% Ag/3DOM $La_{0.6}Sr_{0.4}MnO_3$	In situ PMMA templating	PMMA/ (DMOTEG-PEG400)	300 °C, N_2 (50 ml/min), 3 h → 300 °C, air (50 ml/min), 1 h → 800 °C, air (50 ml/min), 4 h
5.71 wt% Ag/3DOM $La_{0.6}Sr_{0.4}MnO_3$	In situ PMMA templating	PMMA/ (DMOTEG-PEG400)	300 °C, N_2 (50 ml/min), 3 h → 300 °C, air (50 ml/min), 1 h → 800 °C, air (50 ml/min), 4 h

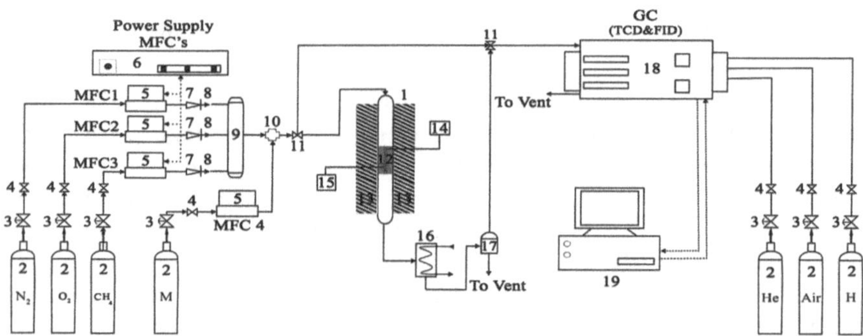

Fig. 2.4 Schematic diagram of the experimental setup: *1* fixed-bed reactor, *2* gas cylinder, *3* regulator, *4* on–off valve, *5* mass-flow controller, *6* power supply MFCs, *7* check valve, *8* needle valve, *9* mixture, *10* cross-flow section, *11* three- and four-way valves, *12* catalyst bed, *13* electrical oven, *14* temperature controller, *15* temperature indicator, *16* condenser, *17* liquid trap, *18* gas chromatograph, *19* computer, and *M* gas mixture

2.2.2 Gas Chromatography

The concentrations of the reactants and products were online-monitored by a GC (Agilent 7890A) equipped with FID and TCD detectors. Using the columns of Porapak-Q/molecular sieve 5A (2 m in length) and RT-QPlot divinylbenzene PLOT (30 m in length). The catalytic performance of the samples was evaluated using the temperatures (T_{10}, T_{50}, and T_{90}) required for methane conversions of 10, 50, and 90 %, respectively. CH_4 conversion was given by $(c_{inlet} - c_{outlet})/c_{inlet} \times 100$ %, where the c_{inlet} and c_{outlet} were the CH_4 concentrations of inlet and outlet feed stream, respectively.

2.3 Further Characterization for Catalyst Tests

All of the catalysts were characterized by means of techniques, such as N_2 adsorption–desorption (Brunauer–Emmett–Teller, BET), X-ray diffraction (XRD), inductively coupled plasma atomic emission spectroscopy (ICP-AES), scanning electron microscopy (SEM), energy-dispersive spectroscopy (EDS), H_2 temperature-programmed reduction (H_2-TPR), thermogravimetric analysis (TGA), transmission electron microscopy (TEM), selected-area electron diffraction (SAED), X-ray photoelectron spectroscopy (XPS), and Fourier transform infrared (FT-IR). The detailed procedures are described in the following sections.

2.3.1 X-Ray Diffraction (XRD) Pattern

The X-ray diffraction (XRD) experiments were carried out on a Philips PW 1800 diffractometer by means of Cu Kα radiation (λ = 0.15406 nm) at 40 kV and 30 mA to conclude the crystalline phases and to analyze the lattice parameters. Scattering intensities were planned over a pointed range of $8° < 2\theta < 90°$ for all of the samples with a step size (2θ) of 0.03° and a count time of 2 s per step. The diffraction patterns were indexed by JCPDS (Joint Committee on Powder Diffraction Standards) files.

2.3.2 Scanning Electron Microscopy (SEM)

The morphologies of the as-prepared samples were considered by scanning electron microscopy (SEM) using a Philips XL30 microscope in use at an accelerating voltage of 30 kV, with a work distance (WD) between 10 and 13 mm and magnification values in the range 100–100,000×. SEM was used in the backscattered electron detector (BSED). The secondary electron detector (SED) modes show the morphologies of the surface structures.

2.3.3 Energy-Dispersive Spectroscopy (EDS)

For the determination of the chemical composition of the crystalline phases, energy-dispersive spectroscopy (EDS) was used to conclude the EDS spectra by means of an EDS DX-4 analysis system. A very thin Au deposit was used to get better the conductivity of the sample such as PMMA.

2.3.4 Transmission Electron Microscopy (TEM) and Selected-Area Electron Diffraction (SAED)

Transmission electron microscopic (TEM) images as well as the selected-area electron diffraction (SAED) patterns of the typical catalysts were verified using a JEOL JEM-2010 apparatus.

2.3.5 BET Surface Area and N_2 Adsorption–Desorption

The surface areas and pore size distributions were planned using the Brunauer–Emmett–Teller (BET) and Barrett–Joyner–Halenda (BJH) methods. The N_2 adsorption–desorption isotherms, surface areas, and pore parameters of the samples were determined via N_2 adsorption at −196 °C on a Micromeritics ASAP 2020 adsorption analyzer. Before measurement, the samples were degassed at 250 °C for 3 h.

2.3.6 Inductively Coupled Plasma-Atomic Emission Spectroscopy (ICP-AES)

For the ICP-AES analyses, the freshly prepared 20 mg samples were introduced into a 60-ml Teflon-made digestion vessel (Savillex, USA). The sample mixed with a solution mixture containing 4 ml (37 % HCl 3 ml + 70 % HNO_3 1 ml), 0.5 ml HF (50 %), and 0.5 ml $HClO_4$ (70 %). The vessel digested at 150 °C with a pressure of 80 psi until a yellowish liquid solution observed. After cooling the vessel to RT, 1 % HCl was added to make sample measurements using OPTIMA 3300 DV model (PerkinElmer, USA).

2.3.7 H_2 Temperature-Programmed Reduction (H_2-TPR)

Hydrogen temperature-programmed reduction (H_2-TPR) studies were conducted with an Autochem II 2920 (Micromeritics) apparatus equipped with a TCD detector. The experiments were conducted on the samples (approximately 50 mg). The samples were firstly flushed with He at a flow rate of 40 ml/min as the heat was increased at a ramp rate of 10 °C/min to 200 °C and held at this temperature for 30 min to remove H_2O. Then, the reducing gas (5.1 % H_2 in Ar) was introduced at a flow rate of 40 ml/min with a ramp rate of 10 °C/min from RT to 750 °C. The difference in H_2 concentration of the effluent was monitored online by the chemical adsorption analyzer. The reduction peak was calibrated against that of complete reduction of a known standard of powdered CuO (Aldrich, 99.995 %).

2.3.8 Temperature-Programmed Reduction of Methane (CH_4 TPR-MS)

Temperature-programmed reduction of methane (CH_4-TPR) was carried out in a continuous-flow quartz microreactor. The sample (100 mg), held in the quartz tube, was degassed in a Helium flow (30 ml/min) at 600 °C for 1 h, and then, the temperature was allowed to decrease to room temperature. The TPR test was conducted in a 20 ml/min mixture of CH_4/He (5 % CH_4) at a ramp rate of 10 °C/min from room temperature to 1000 °C and maintained at this temperature for 100 min. Gases evolved from the microreactor were analyzed with a Pfeiffer Omnistar quadrupolar mass spectrometer.

References

1. Arandiyan HR, Parvari M. Preparation of La-Mo-V mixed-oxide systems and their application in the direct synthesis of acetic acid. J Nat Gas Chem. 2008;17(3):213–24.
2. Sadakane M, Horiuchi T, Kato N, Takahashi C, Ueda W. Facile preparation of three-dimensionally ordered macroporous alumina, iron oxide, chromium oxide, manganese oxide, and their mixed-metal oxides with high porosity. Chem Mater. 2007;19(23):5779–85.
3. Li F, Josephson DP, Stein A. Colloidal assembly: the road from particles to colloidal molecules and crystals. Angew Chem Int Ed. 2011;50(2):360–88.
4. Li F, Qian Y, Stein A. Template-directed synthesis and organization of shaped oxide/phosphate nanoparticles. Chem Mater. 2010;22(10):3226–35.
5. Liu B, Li C, Zhang Y, et al. Investigation of catalytic mechanism of formaldehyde oxidation over three-dimensionally ordered macroporous Au/CeO_2 catalyst. Appl Catal B. 2012;111(1):467–75.
6. Stein A, Wilson BE, Rudisill SG. Design and functionality of colloidal-crystal-templated materials-chemical applications of inverse opals. Chem Soc Rev. 2013;42:2763–803.
7. Stein A, Schroden RC. Colloidal crystal templating of three-dimensionally ordered macroporous solids: materials for photonics and beyond. Curr Opin Solid State Mater Sci. 2001;5(6):553–64.
8. Shen C, Hui C, Yang T, et al. Monodisperse noble-metal nanoparticles and their surface enhanced raman scattering properties. Chem Mater. 2008;20(22):6939–44.
9. Sen T, Tiddy GJT, Casci JL, Anderson MW. Synthesis and characterization of hierarchically ordered porous silica materials. Chem Mater. 2004;16(11):2044–54.
10. Scirè S, Liotta LF. Supported gold catalysts for the total oxidation of volatile organic compounds. Appl Catal B. 2012;125(1):222–46.

11. Deng J, Dai H, Jiang H, et al. Hydrothermal fabrication and catalytic properties of $La_{1-x}Sr_xM_{1-y}Fe_yO_3$ (M = Mn, Co) that are highly active for the removal of toluene. Environ Sci Technol. 2010;44(7):2618–23.
12. Deng J, Zhang L, Dai H, He H, Au CT. Strontium-doped lanthanum cobaltite and manganite: highly active catalysts for toluene complete oxidation. Ind Eng Chem Res. 2008;47(21):8175–83.
13. Deng J, Zhang L, Dai H, He H, Au CT. Hydrothermally fabricated single-crystalline strontium-substituted lanthanum manganite microcubes for the catalytic combustion of toluene. J Mol Catal A: Chem. 2009;299(1):60–7.
14. Deng J, Zhang Y, Dai H, Zhang L, He H, Au CT. Effect of hydrothermal treatment temperature on the catalytic performance of single-crystalline $La_{0.5}Sr_{0.5}MnO_{3-\delta}$ microcubes for the combustion of toluene. Catal Today. 2008;139(1):82–7.

Chapter 3
Performance of the 1D LSCO Nanowires for Methane Combustion

3.1 Introduction

Catalytic combustion of hydrocarbons is a useful strategy for both energy manufacture and environmental pollution abatement [1–4]. In the earlier period, researchers have been running on characterizations and applications of perovskite-type oxides (PTOs) [5]. Among the PTOs materials, those of the La–Sr–Co family have been investigated intensively for catalytic applications [5]. The La–Sr–Co compounds show high level of thermal stability and oxygen mobility, and display stable oxidation states of the La and Co elements [5]. The PTOs catalysts reported in the literature are typically polycrystalline with applicable at low and middling temperatures, since high temperature brings about to reduced surface area below about 2 m^2/g, and it is rare to come across 1D nanomaterials. However, it exhibit abnormal properties such as restricted size, high surface area, peculiar morphology, and novel physical properties and play an important role in fundamental research as well as practical applications [6, 7]. It has been well acknowledged that the morphology of inorganic nanoparticles is a main factor influencing the physics and chemical property of a material. Appreciation for the great synthetic flexibility and controllability of crystalline growth, composite oxide nanowires, and/or nanorods with enhance catalytic activity such as $La_{1-x}Ba_xMnO_3$ [8], 15 % $Ni/MgAl_2O_4$ [9], and $La_{0.6}Sr_{0.4}Fe_{0.8}Bi_{0.2}O_3$ [10]. They have been synthesized by sol–gel method of hard-templating, and the size and shape of the products are frequently determined by the adopted template. Preparation of PTOs and related compounds has been achieved by many procedures, including solid-state reaction [11], mechanochemical solid reaction [12], sol–gel [13], combustion synthesis [14], hydrothermal synthesis [15], sonochemical synthesis [16], thermal decomposition of the heteronuclear complex [17], wet chemical coprecipitation [18], a polymerize able complex method [19], and a microemulsion method [20]. Although there are many studies of preparation of nanosized PTOs thin films and powders [18, 19],

H. Arandiyan, *Methane Combustion over Lanthanum-based Perovskite Mixed Oxides*, Springer Theses, DOI 10.1007/978-3-662-46991-0_3

few papers have been concerned with the preparation and structural properties of PTOs nanowires. For most applications, the controlled synthesis of homogeneous, high-purity and high surface area PTOs materials is necessary for obtaining reproducible properties.

According to this idea, we have lately developed a novel and facile one-pot hydrothermal approach that develops 1D single-crystalline perovskite $La_{0.5}Sr_{0.5}CoO_3$ nanowires without the use of a template, which may be different from those of the nanoparticles. In this chapter, we report the synthesis, characterization, and catalytic performance of nanowire PTOs with dextrose/L-Lysine-assisted hydrothermal route (DHR) for high performance of methane combustion.

3.2 Catalyst Characterization of PTOs

3.2.1 ICP-AES Results

The chemical composition (wt%) of the calcined samples was calculated by ICP-AES. The results specify that the concentration of La, Sr, and Co for the samples was close to the nominal value. The experimental values of PTOs series are, within experimental error, very similar to the theoretical values, indicating a close coincidence between experimental and nominal molecular formula for the PTOs assessing the benefits of the synthesis process to produce highly crystalline and homogeneous catalyst.

3.2.2 XRD Patterns of the Oxides

Taking into consideration, the results obtained by XRD analysis (Fig. 3.1). It can be accomplished that the crystallinity evolution of $La_{0.5}Sr_{0.5}CoO_3$ precursor is a function of the research procedure. The configuration procedure of the perovskite phase can be illustrated at 800 °C. The diffractogram of LSCO-2 at 650 °C reveals that the precursor was totally amorphous. At higher temperatures, an increase in the diffraction line intensity is obtained. The XRD diffractograms of the LSCO-2 illustrate peak with positions and relative intensities regular with those of cubic $La_{0.5}Sr_{0.5}CoO_3$ (JCPDS No.: 48-0122). The diffraction peaks can be well indexed as illustrated in Fig. 3.1 (LSCO-2). The other samples (LSCO-1, LSCO-3, and LSCO-4) have major cubic perovskite phases, but contained trace amounts of impurity phases. It is shown that the quite weak peak at 2θ = ca. 25.3° is attributable to the phase of $SrCO_3$ (JCPDS No.: 05-0418) and/or La_2O_3 (JCPDS No.: 83-1355) that existed as impurity. When the concentration of dextrose in the precursor solution greater than before, the intensity of the 25.3° peak drops off. It appears that the as-obtained catalysts enclosed mainly cubic perovskite-type $La_{0.5}Sr_{0.5}CoO_3$ in both LSCO-1 and LSCO-2. On the other hand, LSCO-3 and LSCO-4 obviously contained $SrCO_3$ (2θ = 25.1°, 25.8°, 29.7°, 44.2°, 50.0°, 65.4°) and (Co_3O_4

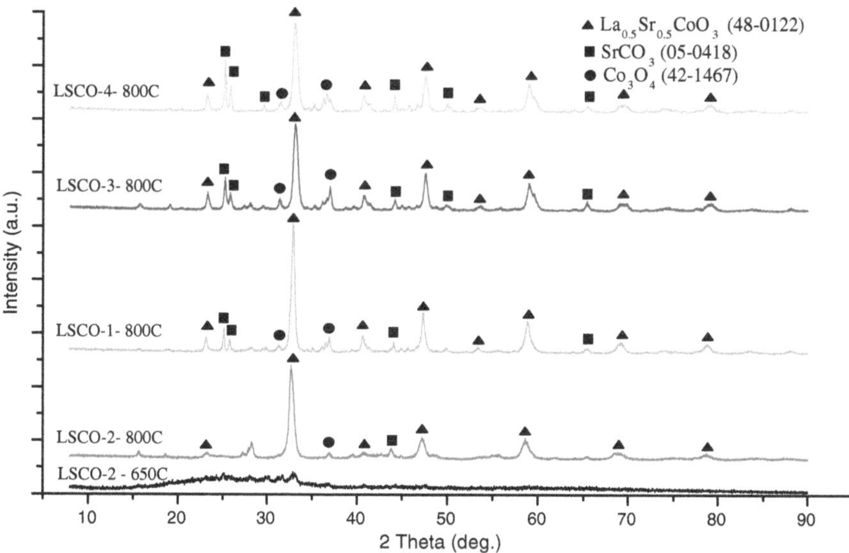

Fig. 3.1 XRD patterns of the oxides (with molar ratio of La/Sr/Co = 0.5/0.5/1.0) prepared by the four different dextrose/L-Lysine volumetric ratio and calcined at 800 °C for 5 h

(2θ = 36.9°) phase in addition to the perovskite main phase (Fig. 3.1). It is recognized that $SrCO_3$ is inactive for the oxidation of organic compounds. Nonetheless, its attendance would undermine the activation of O_2 molecules. In the meantime, it causes a slight reduction in catalytic performance of the PTOs structure. Evidently, the composition of the precursor solution had a great result on the crystal size of the final effect.

3.2.3 SEM/HRSEM Results

Representative (SEM) and (high-resolution SEM) images of LSCO nanowire and nanoparticles are illustrated in Fig. 3.2. Figure 3.2a–c gives details about typical (SEM/HRSEM) LSCO-2 nanowires, when a certain amount of dextrose was added to the precursor solution; however, the as-obtained LSCO-2 samples (except for the LSCO-4 sample (Fig. 3.3d, e) were nanowires with a variable size between 5 and 10 nm (Figs. 3.2 and 3.3). This is outstanding to the fact that the contents of dextrose and L-Lysine applied a large control on the length and width of the individual particle. Although the lengths of nanowires LSCO-2 are less than 2 μm (Fig. 3.2a–c), it does not mean that the nanowires are only so short. Since for the duration of the preparation of catalysts for (SEM) observation, the nanowires were easily broken by the preparation procedure.

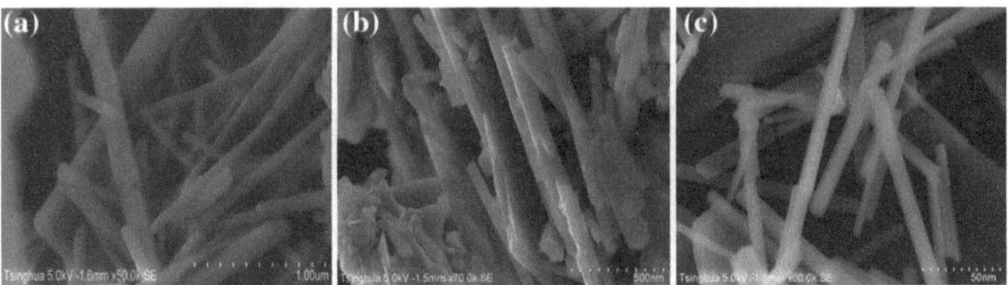

Fig. 3.2 **a**, **b** SEM images; **c** HRSEM image of LSCO-2 nanowires

3.2.4 TEM/HRTEM Results

Figure 3.3 illustrates the high-resolution TEM image of a nanowire LSCO-2. Additionally, the well-resolved distance between the parallel fringes (Fig. 3.3b) was calculated to be ca. 0.103–0.113 nm. It is corresponding to the spacing of [21] planes of LSCO with monoclinic catalyst structure. In order to study the microstructures of the LSCO nanowire and nanoparticles samples, TEM analyses were also conducted, and the typical images of the LSCO-2 and LSCO-4, represented as nanowire and nanoparticles samples, respectively, are shown in Figs. 3.2 and 3.3. The presenting of multiple bright electron diffraction rings selected-area electron diffraction (SAED) patterns (insets of Fig. 3.3a, b).

The diffraction rings are irregular and consist of regularly associated lines of bright spots, which disclosing that the LSCO-2 nanowires/nanorods are well-grown single crystallites. Figure 3.3d, e demonstrates TEM images of nanoparticles LSCO-4 after and before running at 800 °C for 5 h. After experiment, the nanoparticle sample kept the perovskite formation, but their particle size increased. This means that the nanoparticles sintering strictly. Nevertheless, $La_{0.5}Sr_{0.5}CoO_3$ synthesized by dextrose-assisted hydrothermal route (DHR) more or less maintained the nanowire system. This might specify that the nanowires had a higher potential opposed to agglomeration than the nanoparticles.

3.2.5 BET Surface Area

The duration of the samples with diverse morphologies was studied further. Figure 3.4 illustrated the BET surface area variations of the catalysts after and before experiment. Under the reaction conditions of $CH_4/O_2/N_2$ = 10/10/80 ml/min, GHSV = 30,000 ml/(h gcat) at 800 °C for 50 h. For the four LSCO catalysts, it seems that amount of dextrose in the precursor solution was directly relative to the BET of the final product. However, there might be a proper L-Lysin concentration or a most favorable L-Lysine/dextrose volumetric ratio for obtaining the

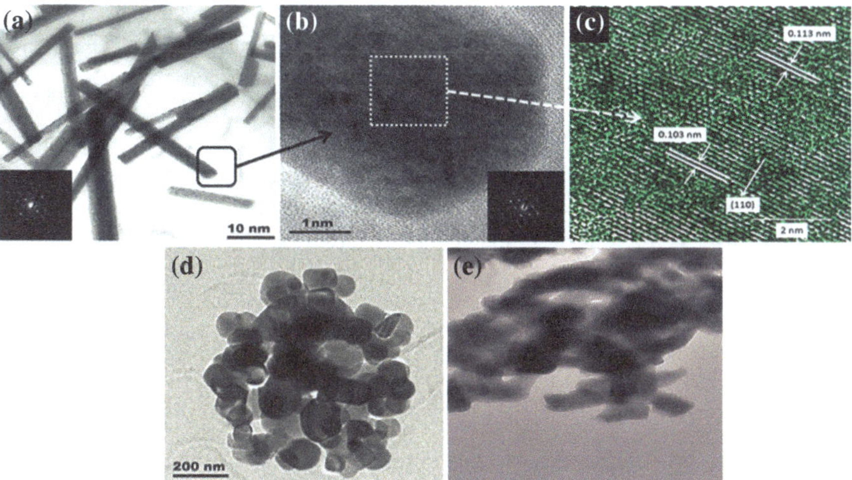

Fig. 3.3 **a** TEM image; **b**, **c** HRTEM images as well as the SAED patterns (*insets*) of (**a** and **b**) of LSCO-2 nanowires; **d** HRTEM image of LSCO-4 nanoparticles before; and **e** after running 800 °C for 50 h under reaction condition: GHSV = 30,000 ml/(h gcat)

maximum BET. In Fig. 3.4, the BET of LSCO-4 catalyst obtained by sol–gel method (19.9 $m^2\ g^{-1}$) before running at 800 °C for 50 h is higher than that of the 800 °C calcined LSCO-2 catalyst obtained by DHR (17.7 $m^2\ g^{-1}$) or even by the LSCO-0, LSCO-1, and LSCO-3 is 4.4, 10.2, 7.3 $m^2\ g^{-1}$, respectively.

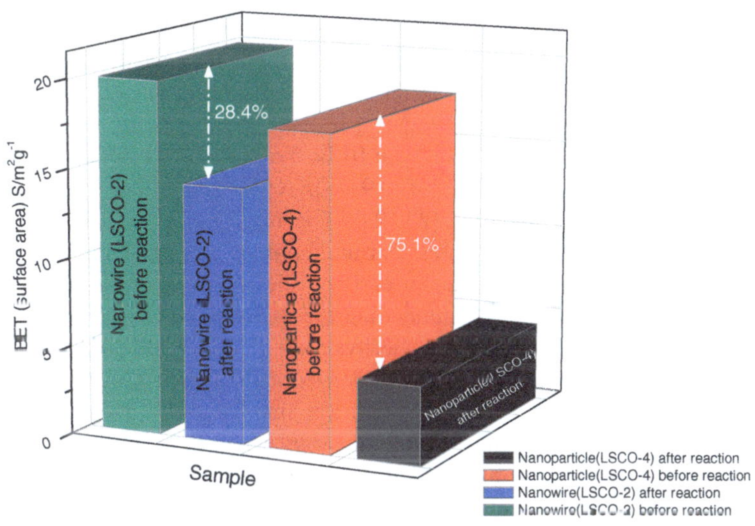

Fig. 3.4 The BET area variations of LSCO catalyst before and after running at 800 °C for 50 h

On the other hand, after running, the BET area of the nanoparticles (LSCO-4) and nanowires (LSCO-2) is 4.4 and 14.3 $m^2\ g^{-1}$, respectively, as explained in Table 3.1. It can be mentioned that the specific (BET) surface area of the nanoparticle sharply decreased about 75.1 %. However, the BET of the nanowire catalyst drooped almost 28.4 %. We can understand that the nanoparticle agglomerated more strictly than the nanowire during the catalytic performance. It is worth note that the BET of our LSCO catalysts was larger than that (<1.8 m^2/g) of $La_{0.5}Sr_{0.5}MnO_3$ nanoparticles obtained via the traditional hydrothermal synthesis method [22].

3.2.6 N_2 Adsorption/Desorption Isotherm

Figure 3.5a, b illustrated the N_2 adsorption/desorption isotherm and pore-size distribution of the catalyst preparation at different routs. We should mention that all the catalysts have mesoporous structure at different preparation methods. The following order of pore-size distribution moved to higher values: LSCO-4 < LSCO-0 < LSCO-3 < LSCO-1 < LSCO-2. The LSCO-2 dextrose/L-Lysine-assisted hydrothermal sample illustrated an adjusted pore structure, and this could be additional returned by the pore-size distribution demonstrated in Fig. 3.5b. For the LSCO-3, LSCO-2, and LSCO-1 catalysts, there was a strong and critical peak at pore diameter = 2–4 nm (Fig. 3.5b).

These catalysts displayed out of the ordinary pore-size distributions, depending on their various pore structures. In addition, LSCO-0 catalyst showed a broad pore-size distribution (from 10 to 25 nm). In the meantime, the LSCO-4 catalyst illustrated much broader pore-size distributions (from 10 to 40 nm) than the other samples. It should be noted that there might also be the presence of a small amount of micropores in all of the catalysts because the dV/dD cm^3/gnm value dropped with the raise up of pore size below 4 nm (Fig. 3.5b). The N_2 adsorption/desorption isotherms be able to categorized as a type III isotherm with a H_3 hysteresis loop in the relative pressure (p/p_0) range of 0.9–1.0, representing that these catalysts occupied a macroporous structure [23, 24]. The macropores in the bulk catalyst were the collected intergranular pores. For the moment, a H_2 hysteresis loop in the p/p_0 range of 0.2–0.9 was detected in each profile of the catalysts (the LSCO-4 catalyst being the exception), recommending the attendance of mesopores [23, 24]. According to IUPAC categorization, the hysteresis loop is type H_3. This type of hysteresis is typically established on solids consisting of aggregates or agglomerates of particles forming slit-shaped pores, with a nonuniform size and/or shape [23]. The synthesized LSCO have high-thermal resistance, and the mesoporous structure stayed toward higher temperatures. As well, the elements at B-sites (i.e., Co in the present case) have a result on the BET of the perovskite-mixed oxides.

Table 3.1 Preparation parameters, BET surface areas, crystallite sizes (D), pore volume, and average pore sizes of the as-prepared $La_{0.5}Sr_{0.5}CoO_3$ catalysts

Catalyst	XRD result		Multipoint BET surface area (m^2/g)				Pore volume (cm^3/g)			Average pore size (nm)[d]
	Crystal phase	D[a] (nm)	Macropore (≥50 nm)	Mesopore (<50 nm)	Total[b]	Total[c]	Macropore (≥50 nm)	Mesopore (<50 nm)	Total	
LSCO-0	Cubic $La_{0.5}Sr_{0.5}CoO_3$ + trace $SrCO_3$	117.2	0.8	10.5	3.6	–	0.138	0.009	0.147	10.3
LSCO-1	Cubic $La_{0.5}Sr_{0.5}CoO_3$ + trace La_2O_3	85.8	1.4	8.8	10.2	–	0.143	0.010	0.153	11.6
LSCO-2	Cubic $La_{0.5}Sr_{0.5}CoO_3$	64.1	2.1	15.6	17.7	14.3	0.231	0.007	0.238	27.4
LSCO-3	Cubic $La_{0.5}Sr_{0.5}CoO_3$ + trace Co_3O_4	40.7	0.9	6.4	7.3	–	0.115	0.006	0.121	16.2
LSCO-4	Cubic $La_{0.5}Sr_{0.5}CoO_3$ + trace $SrCO_3$	53.2	–	–	19.9	4.2	–	–	0.050	26.3

Dex Dextore

[a]The data determined based on XRD results according to the Scherrer's equation

[b]The data were calculated according to the total BET surface area before reactivity test

[c]The data were calculated according to the total BET surface area after reactivity test

[d]The data were calculated according to the BJH method

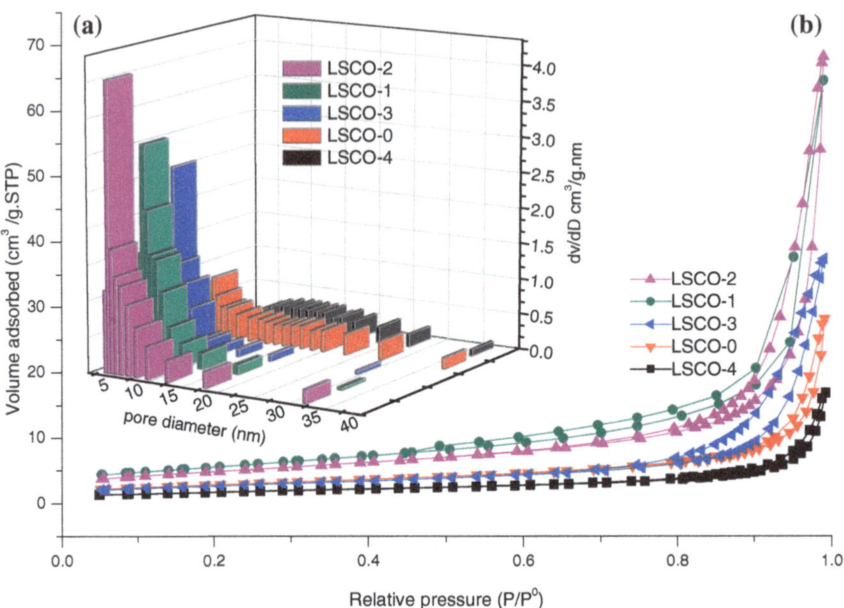

Fig. 3.5 **a** Pore-size distributions; **b** N_2 adsorption/desorption isotherms of $La_{0.5}Sr_{0.5}CoO_3$ synthesized at different methods

3.2.7 Reducibility of Catalysts H_2-TPR Profiles

Figure 3.6 illustrates the TPR profiles of the $La_{0.5}Sr_{0.5}CoO_3$ catalyst at diverse L-Lysine/dextrose ratio. In Fig. 3.6, it is shown that there were big reduction bands at 420 and 801 °C (with a shoulder at 714 °C) over LSCO-1. The corresponding amount of H_2 consumed was 1.34 and 3.45 mmol g cat^{-1}; over LSCO-3, there appeared a big reduction band at 387 °C, a very low reduction band at 521 °C, one critical reduction band at 811 °C, and a shoulder at 714 °C (Fig. 3.6). The total amount of H_2 consumption was 2.57 mmol g cat^{-1} for the former two bands and 4.62 mmol g cat^{-1} for the latter two bands; over LSCO-0, two reduction bands were recorded at 424 and 803 °C (with a shoulder at 718 °C) (Fig. 3.6), corresponding to the H_2 consumption of 1.45 and 3.61 mmol g cat^{-1}.

As can be shown in Fig. 3.6, the use of dextrose/L-Lysine (volumetric ratio of 0.3/1.0) for preparation of LSCO-2 lattice gave rise to four reduction bands at 342 °C (very weak), 419 °C (strong), 643 °C (weak), and 813 °C (very strong); the former two correspond to the H_2 consumption of 2.25 mmol g cat^{-1}, whereas the latter two correspond to that of 4.91 mmol g cat^{-1}. The bands below 500 °C were due to the reduction of O_2 adspecies and the partial reduction of Co^{3+} to Co^{2+}, whereas the ones above 500 °C were due to the reduction of the Co^{2+} [25].

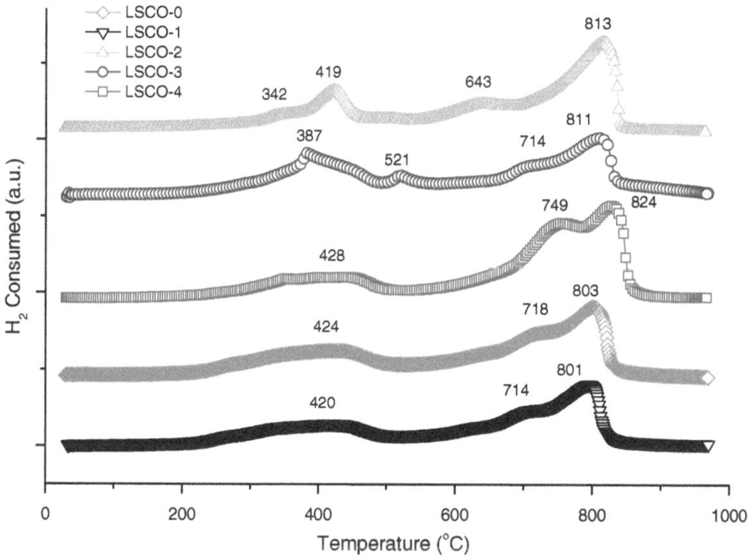

Fig. 3.6 H_2-TPR profiles of $La_{0.5}Sr_{0.5}CoO_3$ catalyst at different preparation methods

3.2.8 Oxygen Species O_2-TPD Profiles

Figure 3.7 shows the formation of the O_2 desorbed during the TPD-O_2 technique at different L-Lysine/dextrose ratio. According to earlier literatures [26], the TPD-O_2 profiles of $La_{0.5}Sr_{0.5}CoO_3$ showed two O_2 desorption peaks. The performance of TPD-O_2 profiles suggests that the peak at low temperatures is associated with adsorbed O_2 (α type), while the other one at high temperatures is fitting to lattice O_2 (β type). The low-temperature peak is, however, less important for CH_4 combustion, since the reaction takes place at much higher temperature. The low-temperature peak has been approved to surface adsorbed O_2, while the O_2 released around 800 °C may be measured as coming from the bulk of the structure. From Fig. 3.7, we can see that the LSCO catalysts have two peaks at about 800 °C (weak) and 900 °C (strong) linked with β-type O_2. It seems that in difference with the XPS results in previews literatures. This disagreement almost certainly arises from catalyst prebehavior used in TPD experiments. The pretreatment in an O_2/Ar flow up to 800 °C prior to record TPD-O_2 profiles look likes to take out the weakly adsorbed O_2 species. It is obvious that the preliminary consumption rate of the samples reduced in the order of LSCO-4 < LSCO-0 < LSCO-3 < LSCO-1 < LSCO-2. As shown in the next section, the catalytic activity of the single-crystalline PTOs pursued a similar order.

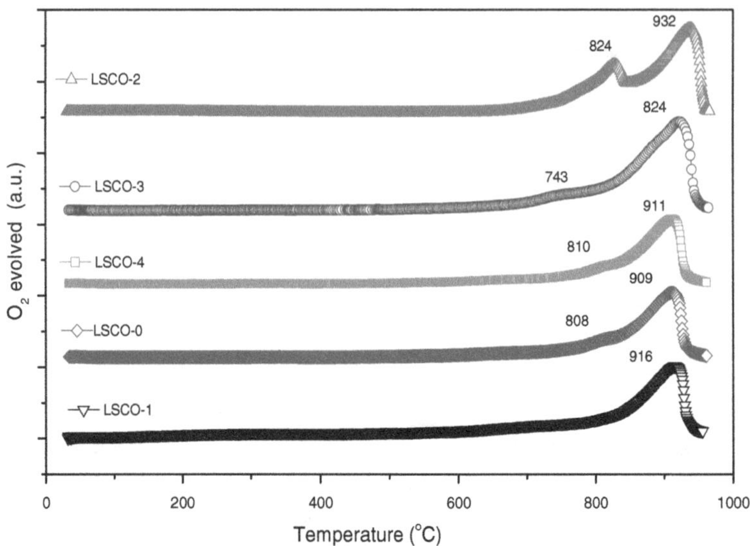

Fig. 3.7 O_2-TPD profiles $La_{0.5}Sr_{0.5}CoO_3$ catalyst at different preparation methods

3.3 Activity Evaluation of Catalyst

3.3.1 Influence of Preparation Method on Catalytic Activity

The performance of CH_4 combustion over the $La_{0.5}Sr_{0.5}CoO_3$ sample synthesize by the four various ratio of L-Lysine/dextrose via hydrothermal method. In addition, one traditional route (sol–gel) stated above as summarized in Table 3.1 calcined at 800 °C for 4 h (shown in Fig. 3.8). The methane conversion raises with increasing reaction temperature; it is understandable that the LSCO-2 and LSCO-4 catalysts achieve the high and low performance activity, respectively. The performance shows that all the samples display 100 % selectivity toward CO_2, and no CO was observed during the reaction. In the meantime, no products of incomplete combustion were detected over our LSCO samples in each run. In the blank test, only quartz sands were loaded in the microreactor. No major conversion of CH_4 was observed below 450 °C, representing that there was no amount of homogeneous reactions over the accepted reaction conditions. Before running at 800 °C for 50 h, the temperatures at $T_{10\%}$, $T_{50\%}$, and $T_{90\%}$ conversion of CH_4 of LSCO-2 as nanowire are ca. 249, 461, and 702 °C. As we can see in Fig. 3.8, they are 101, 61, and 33 °C lower than those (350, 522 and 735 °C) of LSCO-4 as nanoparticle, respectively. The higher performance of methane combustion on nanowire (LSCO-2) can be attributed to its larger BET, higher pore size, and pore volume than those of nanoparticles (LSCO-4). Nanowire catalyst with higher BET may offer more active sites than nanoparticle.

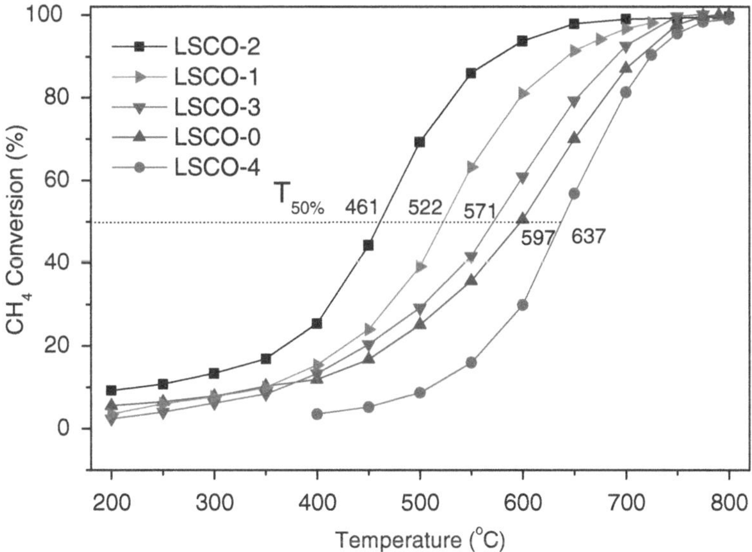

Fig. 3.8 Influence of preparation method on catalytic activity for CH_4 combustion

3.3.2 Influence of Stability and Calcinations at Different Temperatures

As our findings, nanowire LSCO-2 demonstrates a larger catalytic performance than the latter. After running at 800 °C for 3000 min, the conversion of methane on the $La_{0.5}Sr_{0.5}CoO_3$ nanowire (LSCO-2) was maintained (99 %). However, as it can be seen in Fig. 3.9, the LSCO-1, LSCO-3, LSCO-0, and LSCO-4 routes reduced from 98.2 to 84.7, 80.8, 78.1, and 61.4, respectively. After the reaction was finalized, the BET of the nanowire (LSCO-2) reduced a little to 14.4 m^2 g^{-1}. However, nanoparticles (LSCO-4) reduced considerably to 4.4 m^2 g^{-1} (Fig. 3.4). It is understandable that the single-crystal nanowire (LSCO-2) had advanced thermal stability than the nanoparticles (LSCO-4). Several related reasons should be considered. First, the contents of L-Lysine and dextrose are valuable for large effect on catalytic activity, and second, the surface-to-volume ratio of the wires (LSCO-2) is bigger than that of the nanoparticles (LSCO-4). Besides, there is a smaller driving force to agglomerate. Third, the contact area between the wires may be minor, which may also delay agglomerating of the wires (LSCO 2). As a final point, the ideal single crystal structure may also be a factor of the stability. It is well acknowledged that at low conversions of methane, CH_4 combustion is managed mostly by surface catalytic performance and oxygen adspecies [26, 27]. At high conversions of methane, the combustion of CH_4 typically comprises surface reaction and free radical reaction [21]. The free radical reactions are more dependent on mass transfer than on the surface reaction [21]. Before the gases reacted over solid surface, the gas dispersion from bulk gases to the solid surface happened. For the reason that the surface

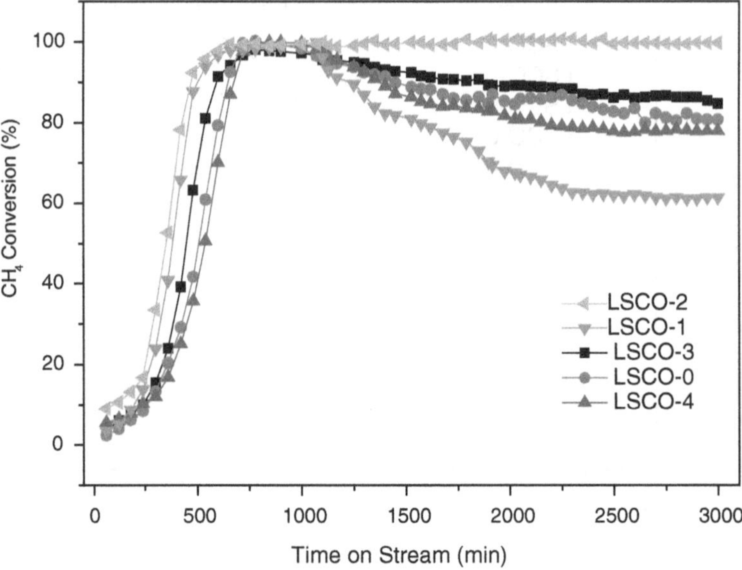

Fig. 3.9 Catalytic lifetime test of different catalysts calcined at 800 °C for 5 h

reaction is fast, the effect of mass transfer from gas bulk to the surface of reaction should not be forgotten. The high BET of the sample is advantageous to mass transfer, because the high BET may support for the adsorption of supplementary gases on the solid surface. As a result, performance of the five oxides for methane combustion is in order of LSCO-4 < LSCO-0 < LSCO-3 < LSCO-1 < LSCO-2.

In order to be able to examine the result of calcination temperature, the precursor with molar ratio of La/Sr/Co = 0.5/0.5/1.0 synthesized calcined at 800 °C for 4 h by the nanowire DHR method. According to XRD patterns, a single phase of mixed oxide perovskite was formed in the calcination at 800 °C for 4 h. Performance activity for methane combustion of the perovskite calcined at the three temperatures is shown in Fig. 3.10. It is understandable that higher temperature calcination bring low CH_4 conversion.

3.3.3 Effects of Space Velocity on the Activity of Catalyst

The methane conversion versus temperature at four different GHSVs (30,000, 35,000, 40,000, and 45,000 ml g cat^{-1} h^{-1}) is shown in Fig. 3.11. The combustion of methane illustrated that the conversion curve moved to lower temperatures at lesser GHSV. The achieved findings illustrated that for all LSCO samples, increasing the space velocities brings a little reduction in methane conversions. A high space velocity can reduce the degree of metallic agglomeration and reduce the size of the crystallites during the reaction [23]. A high space velocity supplies more frequent

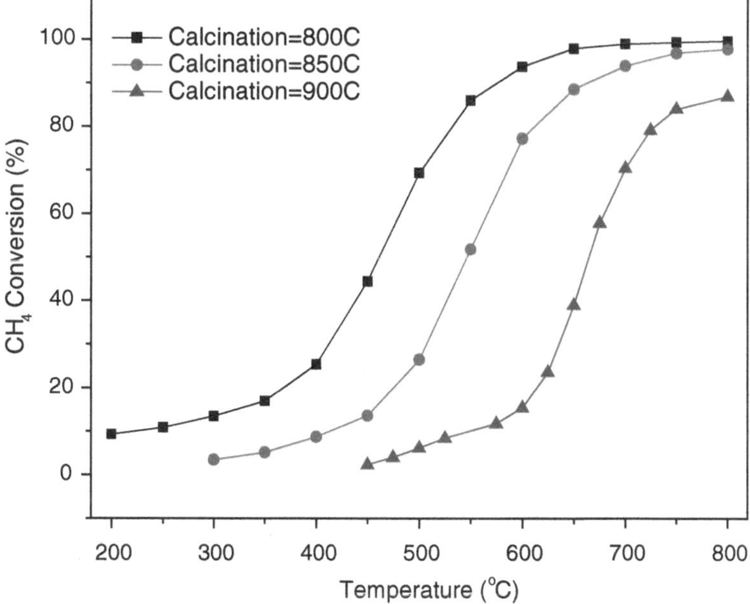

Fig. 3.10 Activities of La/Sr/Co = 0.5/0.5/1.0 after calcinations at different temperatures

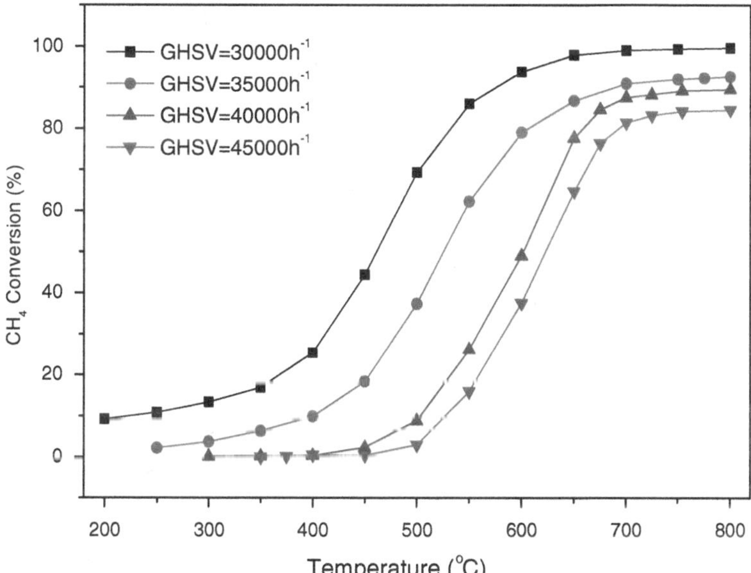

Fig. 3.11 Influence of GHSV on catalytic activity for methane

contact among the sample, not only the reactants (CH_4 and O_2), but also a rapid residence time. The short residence time is an important issue that brings a reduction in methane and oxygen conversions; however, the highest conversion of methane appears at space velocity of 30,000 ml g cat^{-1} h^{-1}. Based on the results of performance tests and H_2-TPR experiments, we believe that the sample with a higher initial H_2 consumption rate in the 300–700 °C range should show a higher catalytic test for combustion of methane. It has been studied that the existence of oxygen vacancies could result in the support of oxygen diffusion and improvement in the mobility of lattice O^{2-} [26, 28]. As shown in Fig. 3.7, O_2 vacancies on/in $La_{0.5}Sr_{0.5}CoO_3$ followed the similar order to the sequence of catalytic activity.

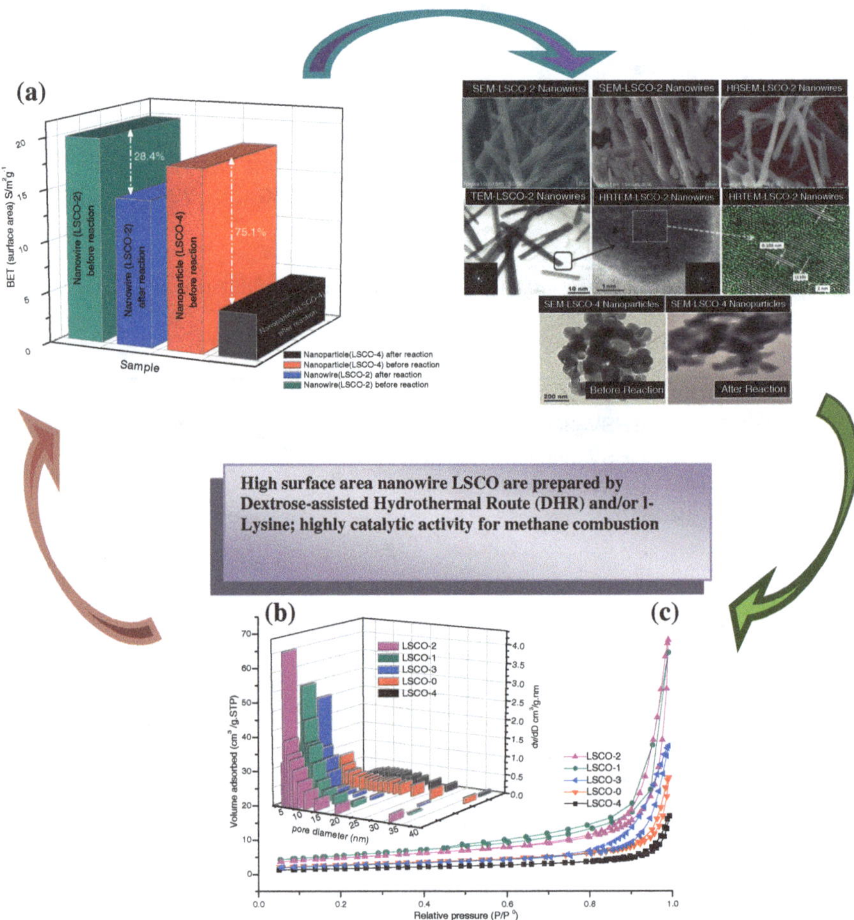

Scheme 3.1 LSCO synthesized via a novel and facile one-pot dextrose-assisted hydrothermal route. High-surface-area nanowire LSCO are prepared by the DHR method. Surface area, O_{ads} content, and reducibility account for the good catalytic activity. Catalytic activity is related to surface area, oxygen adspecies, and reducibility. L-Lysine/dextrose ratio can greatly influence on the morphology and pore structure

3.4 Conclusion and Discussion

In this study, single-crystalline cubic perovskite-mixed oxide $La_{0.5}Sr_{0.5}CoO_3$ nanowire was synthesized via a novel and facile one-pot hydrothermal method with DHR and/or L-Lysine procedure. The volumetric ratio of L-Lysin and dextrose for hydrothermal treatment in the precursor solution applied a large effect on the morphology and pore structure of the $La_{0.5}Sr_{0.5}CoO_3$ samples. The calcination temperature has a considerable impact on the morphology of the $La_{0.5}Sr_{0.5}CoO_3$ materials as well. Between the four synthesized samples, the BET of nanowire LSCO-2 after performance (L-Lysine/glucose volumetric ratio of 1.0/0.3) 14.3 $m^2\ g^{-1}$ is much larger than that of nanoparticle LSCO-4 (4.4 $m^2\ g^{-1}$). Effect on surface property, crystal size, morphology, and also catalytic performance in CH_4 combustion at different temperatures (200–800 °C) has been carefully considered. Under the conditions of 40 ml/min (GHSV = 30,000 ml/(h g_{cat}), $CH_4/O_2/N_2$ = 10/10/80 ml/min), the LSCO-2 catalyst illustrated the high performance, giving the $T_{10\%}$, $T_{50\%}$, and $T_{90\%}$ of ca. 249, 461, and 702 °C, respectively. It is summarized that high O_2 adspecies concentration and excellent low-temperature reducibility as well as unique nanowire morphology were responsible for the best catalytic activity of the LSCO-2 catalyst. We suppose that such nanowire LSCO structure shows potential in CH_4 combustion in real application (Scheme 3.1).

References

1. Auer R, Alifanti M, Delmon B, Thyrion FC. Catalytic combustion of methane in the presence of organic and inorganic compounds over $La_{0.9}Ce_{0.1}CoO_3$ catalyst. Appl Catal B. 2002;39 (4):311–8.
2. Zhong Z, Chen K, Ji Y, Yan Q. Methane combustion over B-site partially substituted perovskite-type $LaFeO_3$ prepared by sol–gel method. Appl Catal A. 1997;156(1):29–41.
3. Machida M, Eguchi K, Arai H. Effect of structural modification on the catalytic property of Mn-substituted hexaaluminates. J Catal. 1990;123(2):477–85.
4. Duprat AM, Alphonse P, Sarda C, Rousset A, Gillot B. Nonstoichiometry-activity relationship in perovskite-like manganites. Mater Chem Phys. 1994;37(1):76–81.
5. Niu J, Deng J, Liu W, et al. Nanosized perovskite-type oxides $La_{1-x}Sr_xMO_{3-\delta}$ for the catalytic removal of ethylacetate. Catal Today. 2007;126(3):420–9.
6. Zhang G, Wang D, Möhwald H. Patterning microsphere surfaces by templating colloidal crystals. Nano Lett. 2004;5(1):143–6.
7. Wang Z, Kiesel ER, Stein A. Silica free syntheses of hierarchically ordered macroporous polymer and carbon monoliths with controllable mesoporosity. J Mater Chem. 2008;18 (19):2194–200.
8. Urban JJ, Ouyang L, Jo MH, Wang DS, Park H. Synthesis of single-crystalline $La_{1-x}Ba_xMnO_3$ nanocubes with adjustable doping levels. Nano Lett. 2004;4(8):1547–50.
9. Ma X, Zhang H, Xu J, et al. Synthesis of $La_{1-x}Ca_xMnO_3$ nanowires by a sol–gel process. Chem Phys Lett. 2002;363(5):579–82.
10. Zhao Z, Dai H, Deng J, Du Y, Liu Y, Zhang L. Preparation of three-dimensionally ordered macroporous $La_{0.6}Sr_{0.4}Fe_{0.8}Bi_{0.2}O_{3-\delta}$ and their excellent catalytic performance for the combustion of toluene. J Mol Catal A Chem. 2013;366(1):116–25.

11. Kakihana M. Invited review "sol-gel" preparation of high temperature superconducting oxides. J Sol–Gel Sci Technol. 1996;6(1):7–55.
12. Zhang Q, Saito F. Effect of Fe_2O_3 crystallite size on its mechanochemical reaction with La_2O_3 to form $LaFeO_3$. J Mater Sci. 2001;36(9):2287–90.
13. Rivas I, Alvarez J, Pietri E, Pérez-Zurita MJ, Goldwasser MR. Perovskite-type oxides in methane dry reforming: effect of their incorporation into a mesoporous SBA-15 silica-host. Catal Today. 2010;149(3):388–93.
14. Qi X, Zhou J, Yue Z, Gui Z, Li L. A simple way to prepare nanosized $LaFeO_3$ powders at room temperature. Ceram Int. 2003;29(3):347–9.
15. Zheng W, Liu R, Peng D, Meng G. Hydrothermal synthesis of $LaFeO_3$ under carbonate-containing medium. Mater Lett. 2000;43(1):19–22.
16. Deng J, Zhang L, Dai H, Au C-T. In situ hydrothermally synthesized mesoporous $LaCoO_3$/SBA-15 catalysts: high activity for the complete oxidation of toluene and ethyl acetate. Appl Catal A. 2009;352(1):43–9.
17. Sadakane M, Takahashi C, Kato N, et al. Three-dimensionally ordered macroporous (3DOM) materials of spinel-type mixed iron oxides. Synthesis, structural characterization, and formation mechanism of inverse opals with a skeleton structure. Bull Chem Soc Jpn. 2007;80(4):677–85.
18. Liang JJ, Weng H-S. Catalytic properties of lanthanum strontium transition metal oxides ($La_{1-x}Sr_xBO_3$; B = manganese, iron, cobalt, nickel) for toluene oxidation. Ind Eng Chem Res. 1993;32(11):2563–72.
19. Kakihana M. Invited review "sol–gel" preparation of high temperature superconducting oxides. J Sol–Gel Sci Technol. 1996;6(1):7–55.
20. Deng J, Zhang L, Dai H, He H, Au C. Single-crystalline nanowires/nanorods derived hydrothermally without the use of a template: catalysts highly active for toluene complete oxidation. Catal Lett. 2008;123(3):294–300.
21. Royer S, Bérubé F, Kaliaguine S. Effect of the synthesis conditions on the redox and catalytic properties in oxidation reactions of $LaCo_{1-x}Fe_xO_3$. Appl Catal A. 2005;282(1):273–84.
22. Deng J, Zhang Y, Dai H, Zhang L, He H, Au CT. Effect of hydrothermal treatment temperature on the catalytic performance of single-crystalline $La_{0.5}Sr_{0.5}MnO_{3-\delta}$ microcubes for the combustion of toluene. Catal Today. 2008;139(1):82–7.
23. Yan H, Blanford CF, Holland BT, Smyrl WH, Stein A. General synthesis of periodic macroporous solids by templated salt precipitation and chemical conversion. Chem Mater. 2000;12(4):1134–41.
24. Li H, Zhang L, Dai H, He H. Facile synthesis and unique physicochemical properties of three-dimensionally ordered macroporous magnesium oxide, gamma-alumina, and ceria–zirconia solid solutions with crystalline mesoporous walls. Inorg Chem. 2009;48(10):4421–34.
25. Araujo GCd, Lima SMd, Assaf JM, Peña MA, Fierro JLG, do Carmo Rangel M, Catalytic evaluation of perovskite-type oxide $LaNi_{1-x}Ru_xO_3$ in methane dry reforming. Catal Today 2008;133(1):129–35.
26. Seiyama T. Total oxidation of hydrocarbons on perovskite oxides. Catal Rev. 1992;34 (4):281–300.
27. Johansson EM, Danielsson KMJ, Pocoroba E, Haralson ED, Järås SG. Catalytic combustion of gasified biomass over hexaaluminate catalysts: influence of palladium loading and ageing. Appl Catal A. 1999;182(1):199–208.
28. Royer S, Van Neste A, Davidson R, McIntyre B, Kaliaguine S. Methane oxidation over nanocrystalline $LaCo_{1-x}Fe_xO_3$: resistance to SO_2 poisoning. Ind Eng Chem Res. 2004;43 (18):5670–80.

Chapter 4
3DOM LSMO with High Surface Areas for the Combustion of Methane

4.1 Introduction

Catalytic combustion of CH_4 is significant for power generation and environment protection. Perovskite-type oxides (ABO_3), A is typically an alkaline earth and B a transition-metal ion [1–6]. Unfortunately, the traditional methods (e.g., solid-state reaction [7], sol–gel [8], coprecipitation [9], and polymerizable complexing [10]) are concerned in high-temperature reaction, bring about the damaging of pore structures, and consequently show low BET (<10 m^2/g). This problem can be resolved by means of colloidal crystal templating (CCT) route, by which one can create a three-dimensionally ordered 3DOM structure. The colloidal crystal template is fabricated by ordering monodispersed microspheres (poly methyl methacrylate (PMMA) or silica) into a face-centered close-packed array [11]. With PS microbeads as template, Sadakane et al. [12–14] fabricated 3DOM $La_{1-x}Sr_xFeO_3$ ($x = 0–0.4$) with a surface area of 24–49 m^2/g. Using PMMA as hard template [15], have obtained 3DOM $BiVO_4$ with a surface area of 18–24 $m^2\ g^{-1}$, $y$$CoO_x$/3DOM $Eu_{0.6}Sr_{0.4}FeO_3$ (y = 1, 3, 6, and 10 wt%) with a surface area of 22–31 m^2/g [16], and 3DOM $La_{0.6}Sr_{0.4}Fe_{0.8}Bi_{0.2}O_3$ with a surface area of 20–27 m^2/g [15].

In previous chapter, we fabricated 1D LSCO nanowire materials with high BET without using the surfactant-templating. In this chapter, we compare result of 1D to 3D catalyst as well as investigate on surfactant-assisted PMMA method. The use of a template can give increase to the invention of porous structure on the wall of skeletons. Note that some of the 3DOM catalysts showed brilliant catalytic activity in the combustion of CH_4, therefore enhancing the physical and chemical properties of the catalysts. Based on this new idea, we have developed a novel and facile dimethoxytetraethylene glycol (DMOTEG) PMMA-templating route, in order to be able to synthesize 3DOM $La_{0.6}Sr_{0.4}MnO_3$ (3DOM LSMO) with nanovoid-like or mesoporous walls. In this chapter, we report the characterization, preparation, and performance activity of CH_4 combustion of 3DOM LSMO with nanovoid-like or

H. Arandiyan, *Methane Combustion over Lanthanum-based Perovskite Mixed Oxides*, Springer Theses, DOI 10.1007/978-3-662-46991-0_4

mesoporous skeletons by the use of the surfactant [DMOTEG, L-lysine, poly (ethylene glycol), (PEG400), triblock copolymer (Pluronic P-123), and/or ethylene glycol (EG)]-assisted PMMA-templating strategy.

4.2 Catalytic Characterization of 3DOM LSMO

4.2.1 Crystal Structure (XRD)

The XRD patterns of the 3DOM LSMO catalysts are illustrated in Fig. 4.1a, b. The characteristic diffraction peaks with 2θ values of 22.9°, 32.6°, 40.2°, 46.8°, 52.9°, 58.1°, 68.6°, and 78.1° matched with the (012), (110), (202), (024), (116), (300), (208), and (128) lattice planes of (JCPDS PDF# 50-0308) the rhombohedral perovskite phase. From the enlarged consider of the XRD patterns in the 2θ range of 26°–38°, one can detect the change in peak intensity (Fig. 4.1b), representing the attendance of discrepancy in crystallinity of these catalysts. The average crystallite sizes of the 3DOM LSMO-DP1, LSMO-DP3, LSMO-DP5, LSMO-LP1, and LSMO-PE1 were ca. 24.8 nm. No noticeable changes in crystallite size were found. Increasing calcination temperature from 800 to 850 or 900 °C, however, the obtained LSMO-DP3-800, LSMO-DP3-850, and LSMO-DP3-900 as well as the 1D LSMO (calcined at 850 °C) catalysts displayed enhanced crystallinity. As can be seen from Table 4.1, the crystallite size (61.4 nm) of 1D LSMO was much larger

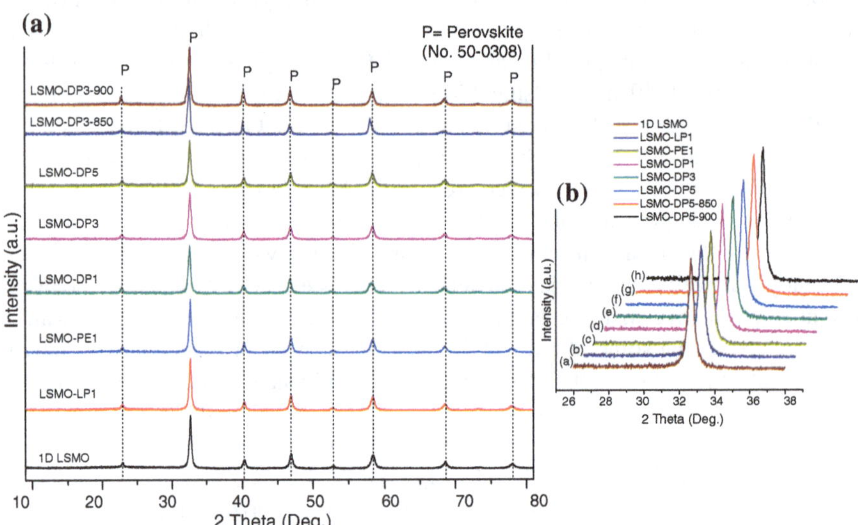

Fig. 4.1 **a** XRD patterns and **b** study on the XRD highest peak of (*a*) 1D LSMO, (*b*) LSMO-LP1, (*c*) LSMO-PE1, (*d*) LSMO-DP1, (*e*) LSMO-DP3, (*f*) LSMO-DP5, (*g*) LSMO-DP5-850, and (*h*) LSMO-DP5-900

Table 4.1 Preparation parameters, BET surface areas, crystallite sizes (D), pore volumes, and average pore sizes of the 3DOM LSMO and 1D LSMO samples

Catalyst code	XRD result		Metal nominal content (%)[b]			BET surface area (m^2/g)			Pore volume (cm^3/g)			Average pore size (nm)[d]
	Crystal phase	D^a (nm)	La	Sr	Mn	Macropore (≥50 nm)	Mesopore (<50 nm)	Total	Macropore (≥50 nm)	Mesopore (<50 nm)	Total	
1D LSMO	Rhombohedral	51.4	12.32	8.71	61.52	–	–	2.6	–	–	–	–
LSMO-DP1	Rhombohedral	26.7	12.39	8.19	61.53	6.4	31.2	37.6	0.125	0.009	0.134	2.2
LSMO-DP3	Rhombohedral	23.5	12.45	8.99	59.96	7.3	34.8	42.1 (38.2)[c]	0.151	0.008	0.159	3.4
LSMO-DP5	Rhombohedral	24.3	–	–	–	5.2	29.9	35.1	0.122	0.006	0.128	2.1
LSMO-LP1	Rhombohedral	24.1	12.13	8.65	61.25	5.1	29.1	34.2	0.119	0.007	0.126	1.6
LSMO-PE1	Rhombohedral	25.6	12.76	8.44	61.43	5.7	25.9	31.6	0.112	0.006	0.118	1.5
LSMO-DP3-850	Rhombohedral	43.7	–	–	–	1.8	15.3	17.1	0.115	0.005	0.120	1.3
LSMO-DP3-900	Rhombohedral	51.2	–	–	–	2.2	5.4	7.6	0.107	0.003	0.110	1.7

[a]The data determined based on the XRD results according to the Scherrer's equation
[b]Chemical compositions determined by the ICP-AES technique
[c]The datum in the parenthesis was determined after 24 h of onstream reaction
[d]The data were calculated according to the BJH method

than those (23.5–26.7 nm) of LSMO-DP1-5, LSMO-LP1, and LSMO-PE1. Furthermore, increasing calcination temperature from 800 to 850 or 900 °C, the crystallite size of LSMO-DP3 enlarged mainly from 23.5 to 43.7 or 51.2 nm.

4.2.2 Scanning Electron Microscopy (SEM)

Figure 4.2 illustrates the SEM and HRSEM images of the LSMO catalysts. It is studied that after calcination at 800 °C, all of the LSMO catalysts shown a 3DOM structure with unlike property (Fig. 4.2a–o). While, the 1D LSMO catalyst illustrated a variable morphology with a particle size of 200–600 nm (Fig. 4.2p). There were amount of 3DOM architectures (pore size = 180–200 nm) in all of the synthesized porous LSMO catalysts (Fig. 4.2a–o). Adding of ethylene glycol (5.00 ml) and Pluronic P-123 (1.20 g) in the existence of MeOH (3.00 ml), a high-property 3DOM architecture was formed in the LSMO-PE1 catalyst (Fig. 4.2j, k). Meanwhile, presence of surfactants PEG400 (3.00 ml) and L-lysine (1.00 g) can enhance property of 3DOM LSMO-LP1 (Fig. 4.2h, i). With the introduction of DMOTEG, high-quality 3DOM structure was produced in the other five LSMO catalysts. We should mention that their macropore characteristic was different in the lead synthesis circumstances (Fig. 4.2a–g, l–o). Increasing the DMOTEG

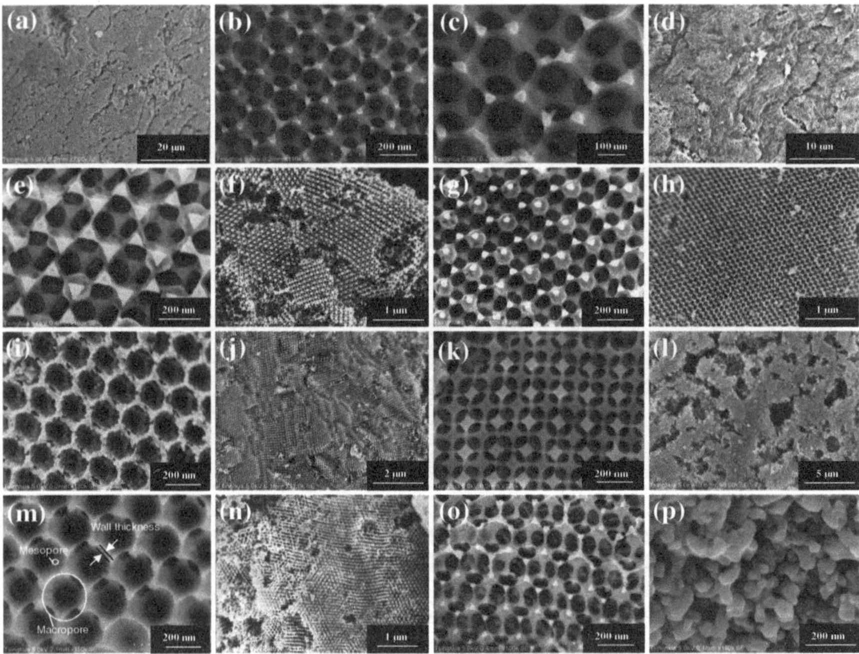

Fig. 4.2 SEM images of **a–c** LSMO-DP3, **d**, **e** LSMO-DP1, **f**, **g** LSMO-DP5, **h**, **i** LSMO-LP1, **j**, **k** LSMO-PE1, **l**, **m** LSMO-DP3-850, **n**, **o** LSMO-DP3-900, and **p** 1D LSMO

concentration from 1.00 to 3.00 ml with PEG400 (5.00 ml) brings the formation of excellent high-property 3DOM architecture with interrelated pore walls in the LSMO-DP3 catalyst (Fig. 4.2a). Increasing the calcination temperature from 800 to 850 or 900 °C, the gained LSMO-DP3-850 and LSMO-DP3-900 catalysts produced in a modification in pore-wall construction. In the meantime, their crystallite sizes (47–72 nm) raised as related to those (27–37 nm) of the 3DOM LSMO catalysts fabrication after calcination at 800 °C. However, the 3DOM arrays were maintained. The addition of the DMOTEG surfactant from 3.00 to 5.00 ml increased the size of LSMO DP5 catalyst. This also showed 3DOM architecture (Fig. 4.2f, g and A.2–A.7 in Appendix). On the contrary, 3DOM property became not as good as compared to that of the LSMO-DP3 catalyst. The macropore diameters of LSMO-DP3, LSMO-LP1, and LSMO-PE1 were between 160 and 210 nm. Meanwhile, average pore sizes of 3DOM LSMO calculated approximately from the HRSEM images were 250–300 nm, corresponding to the reduction of 15–25 % as related to the original size of PMMA microspheres. Thus, the LSMO-DP3 catalyst controlled conventional macropores and uniform average diameters. Besides, wall thickness and macropores were very well regularly arrayed and interconnected through small windows.

4.2.3 Transmission Electron Microscopy (TEM)

An additional obvious issue is the wall structure of the 3DOM catalyst, which could be obviously studied from the TEM and HRTEM images (Fig. 4.3). Surfactant DMOTEG had a vital role to participate in the development of nanovoids on the skeletons of 3DOM LSMO. In the lack of DMOTEG, no or less amounts of nanovoids were formed in the obtained catalysts (Fig. 4.3c, f, j, m, p, t). There were obvious lattice fringes with the lattice spacing (*d* values) of the (110) plane being 0.276–0.277 nm for the LSMO-DP1, LSMO-DP3, and LSMO-DP5 catalysts. Considerably, they are close to 0.2755 nm (JCPDS PDF# 50-0308), and 0.275–0.277 nm for the LSMO-LP1, LSMO-PE1, and LSMO-DP3-900 catalysts, not far away from that of the standard $La_{0.67}Sr_{0.33}MnO_{2.91}$ catalyst. Additionally, the recording of several bright electron diffraction rings in the SAED patterns (insets Fig. 4.3c, f, j, m, p, t) shows that all of the 3DOM-architecture catalysts were polycrystalline. In the meantime, with increasing in calcination temperature, the rise in quantity of electron diffraction rings of LSMO-DP3-850 and LSMO-DP3-900 approves the development in crystallity. The loss of the majority of the nanovoids in the walls of the 3DOM LSMO catalysts (Fig. 4.3q–t) may be connected to the upper calcination temperatures. Extra TEM images of 3DOM-structures LSMO-LP1, LSMO-PE1, and LSMO-DP5 catalysts are illustrated in Fig. A.8 in the Appendix. Through the synthesis procedure of 3DOM LSMO, DMOTEG would function as a multifunctional surfactant for the development of 3DOM structure with brilliant chemical. The DMOTEG-assisted method not only produced high-property 3DOM architecture, but also became stable as it opposed to sintering and

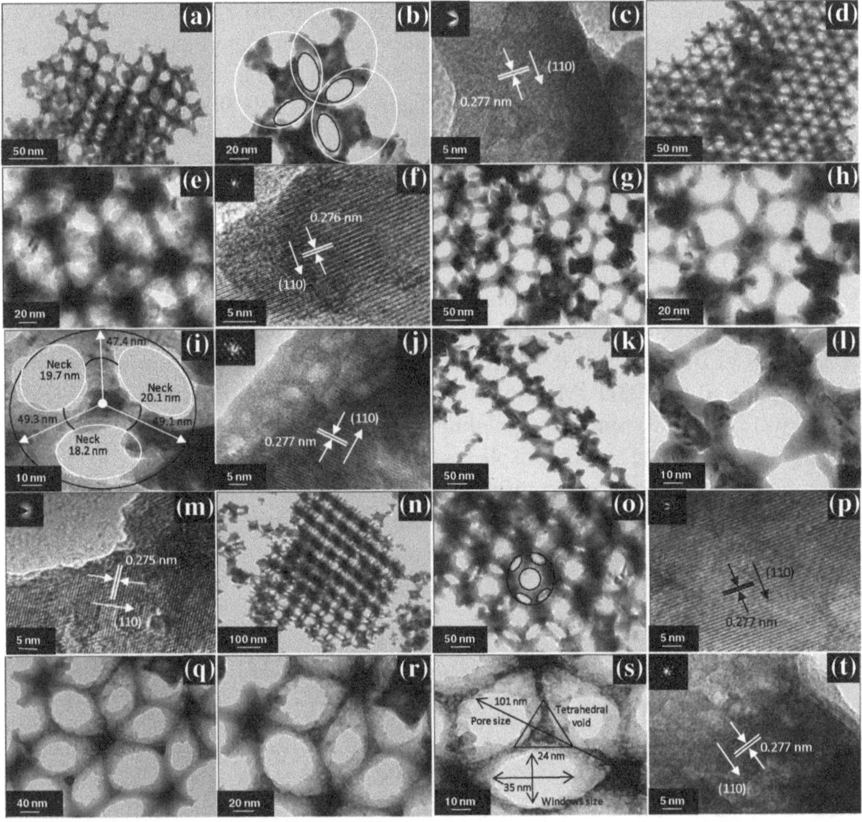

Fig. 4.3 TEM images and SAED patterns (*insets*) of **a–c** LSMO-DP3, **d–f** LSMO-DP1, **g–j** LSMO-DP5, **k–m** LSMO-LP1, **n–p** LSMO-PE1, and **q–t** LSMO-DP3-900

agglomeration. These results show that the nature of reagent and solvent had significant factors on the production of nanovoids in the macropore skeletons of 3DOM LSMO. Thus, macropores had long-range ordered structure, regular hole patterns, and uniform hole walls.

4.2.4 BET Surface Area and N_2 Adsorption/Desorption Isotherms

Figure 4.4 illustrates pore-size distributions and N_2 adsorption–desorption isotherms of the LSMO catalysts, besides their pore parameters as well as BET are summarized in Table 4.1. A type II of N_2 adsorption–desorption isotherm with a type H3 hysteresis loop in the relative pressure (p/p_0) range of 0.9–1.0 was studied for the 3DOM LSMO catalysts.

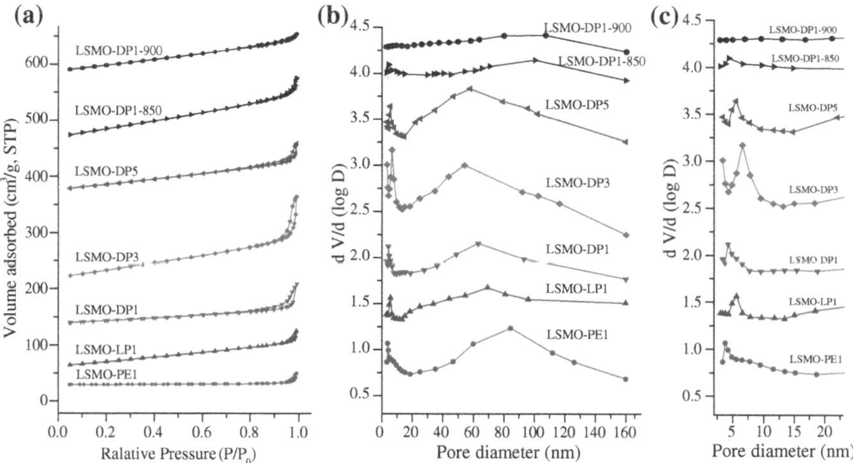

Fig. 4.4 a Nitrogen adsorption–desorption isotherms and **b**, **c** pore-size distributions of the 3DOM LSMO samples

In the low-pressure portion, near linear center section of each of the isotherms was detected, which is attributable to the clear mono-multilayer adsorption, suggesting that the catalysts were nonporous or macroporous adsorbents [16]. The large increase in adsorption at raised relative pressure was characteristics of the slightly assembled walls [17]. For the LSMO-DP1, LSMO-DP3, and LSMO-DP5 catalysts, there was a low peak but with high intensity at pore diameter = 3–6 nm (Fig. 4.4b). Besides, LSMO-DP3-850 and LSMO-DP3-900 catalysts and other catalysts showed a broad pore-size distribution between 10 and 60 nm. The 1D LSMO and LSMO-PE1 catalysts illustrate much broader pore-size distributions between 20 and 90 nm than the other catalysts (Fig. 4.4b, c). From Table 4.1, one can find that a increase in the calcination temperature from 800 to 850 or 900 °C brought a marked drop in BET from 42.1 m^2/g of LSMO-DP3 to 17.1 m^2/g of LSMO-DP3-850, and 7.6 m^2/g of LSMO-DP3-900, a effect probably due to the loss of most of the nanovoids in the walls. It is worth mentioning that LSMO-DP3-850 shows a much larger BET (17.1 m^2/g) than the 1D LSMO catalyst (2.6 m^2/g).

4.2.5 CH_4 TPR-MS Results

The CH_4 activation over the 3DOM LSMO-DP3 sample was examined using the TPR-MS system. The progress of all gases (including CO_2, CH_4, H_2O, and H_2) during the CH_4-TPR test over the 3DOM LSMO-DP3 sample is illustrated in Fig. 4.5.

A whole analysis of all the gases progressed during the CH_4-TPR process gives important hints on desorption of the gases. As can be seen in Fig. 4.5, the complete oxidation of CH_4 ($CH_4 + 4O_{solid} \rightarrow CO_2 + 2H_2O$) prevailed at lower temperatures,

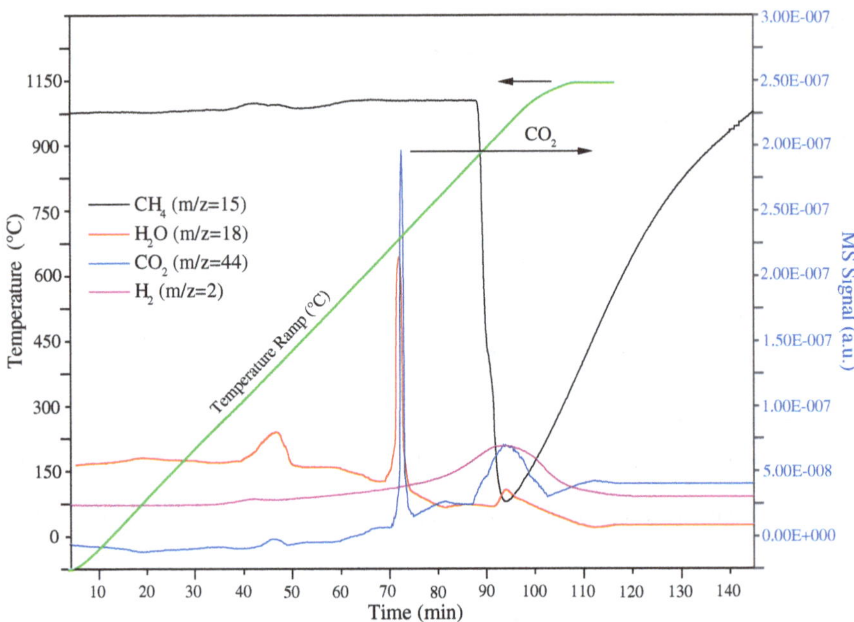

Fig. 4.5 Evolution of the indicated gases [including *m*/*e* = 2 (H_2), 15 (CH_4), 18 (H_2O), and 44 (CO_2)] during the CH_4-TPR test over LSMO-DP3 catalyst

pursued by the partial oxidation of methane ($CH_4 + O_{solid} \rightarrow CO + 2H_2$, $CH_4 + 2O_{solid} \rightarrow CO + 2H_2O$). On the other hand, CH_4 cracking and generation of carbon deposits ($CH_4 \rightarrow C + 2H_2$) probably occurred at high temperatures, which were in good accordance with the literatures [18–20].

4.2.6 Surface Composition, Metal Oxidation State, and Oxygen Species (XPS)

Figure 4.6 illustrates the Mn $2p_{3/2}$ and O 1s XPS spectra of the LSMO catalysts, and surface element compositions are also investigated in Table 4.2. It is shown from Fig. 4.6a that each of the Mn $2p_{3/2}$ spectra was asymmetrical and could be decomposed into three components: surface Mn^{4+} species at binding energy = 642.7 eV, surface Mn^{3+} species at binding energy = 641.9 eV, and the satellite at binding energy = 644.1 eV [5, 21, 22]. Adding reagent DMOTEG, the surface Mn^{4+}/Mn^{3+} molar ratios of the LSMO-DP catalysts raised, and the LSMO-DP3 and LSMO-DP1 catalysts demonstrated larger BET Mn^{4+}/Mn^{3+} molar ratios (1.45 and 1.32, respectively). It is noted that the surface La/Mn (0.92–0.98) and La/(Mn + La) (0.48–0.49) molar ratios of LSMO-PE1 and LSMO-LP1 were lower than the equivalent nominal Mn/La (1.00) and Mn/(Mn + La) (0.50) molar ratios

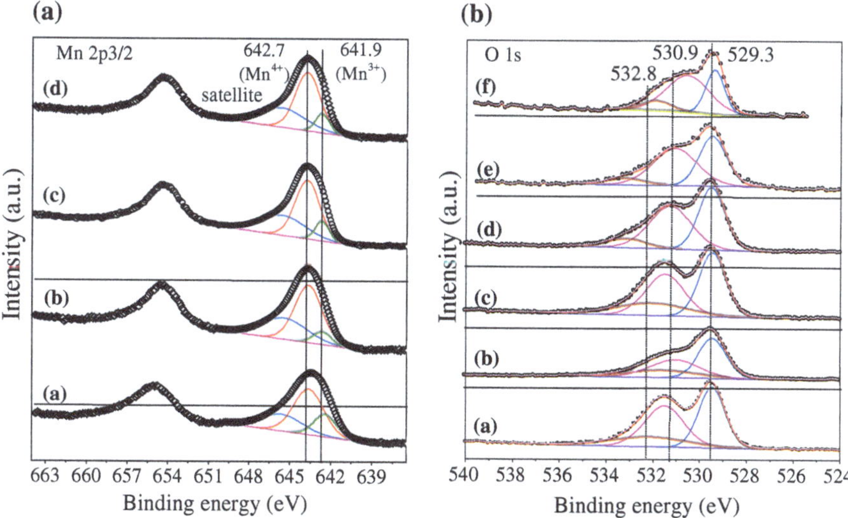

Fig. 4.6 **a** Mn $2p_{3/2}$ and **b** O 1s XPS spectra of (*a*) LSMO-PE1, (*b*) LSMO-DP1, (*c*) LSMO-DP3, (*d*) LSMO-DP5, (*e*) 1D LSMO, and (*f*) LSMO-LP1

(Table 4.2), perhaps the presence of Mn enrichment on the surface of the two catalysts. As shown in Fig. A.9 Appendix, the surface lattice oxygen (O_{lat}) species at binding energy = 529.1–529.3 eV, surface adsorbed oxygen (O_{ads}, O_2^-, O_2^{2-}, O^- or CO_3^{2-}) species at binding energy = 530.9–531.1 eV, as well as surface adsorbed molecular H_2O at binding energy = 532.8 eV [5, 23, 24]. Higher O_{ads} species concentration would be beneficial for the improvement in performance of catalysts for the total oxidation of methane. The surface O_{ads}/O_{lat} molar ratios (Table 4.2) reduced in the order of LSMO-DP3 (1.22) > LSMO-DP1 (1.01) > LSMO-DP5 (0.99) > LSMO-LP1 (0.98) > LSMO-PE1 (0.98) > LSMO-DP3-850 (0.97) > LSMO-DP3-900 (0.96) > 1D LSMO (0.81). It is found that the higher BET favored the rise in O_{ads} species concentration of the LSMO catalyst. With the increase in calcination temperature, there was decline in O_{ads} species concentration; however, the O_{ads} species concentrations of the LSMO-DP3-850 and LSMO-DP3-900 catalysts were still higher than that of the 1D LSMO catalysts.

4.2.7 Reducibility (H_2-TPR)

Figure 4.7a and Table 4.2 show the H_2-TPR profiles of the LSMO catalysts. There were two reduction peaks centered at 498 and 783 °C for the 1D LSMO catalysts, besides 419 and 709 °C for the 3DOM LSMO-DP3 catalyst. For LSMO-DP1 and LSMO-DP5 catalysts, there was a main peak centered at ca. 469 and 474 °C, respectively, with a shoulder at ca. 371 °C. The LSMO-PE1 catalyst presented a major peak centered at ca. 411 °C with a shoulder at ca. 477 °C. After counting the

Table 4.2 Surface element compositions, H_2 consumptions, and catalytic activities of the LSMO samples under the conditions of the reactant mixture = 2 % CH_4 + 20 % O_2 + 78 % N_2 (balance) and the total flow = 41.6 ml/min over 20 mg of the catalyst

Catalyst	La				Mn			
	3d		La/Mn	La/(Mn + La)	$2p_{3/2}$		Mn^{4+}/Mn^{3+}	Mn/(La + Mn + O)
	BE[a]		Molar ratio[b]	Molar ratio	BE		Molar ratio	Molar ratio
1D LSMO	834.4	837.9	1.01 (1.00)	0.51 (0.50)	641.9	642.7	0.91	0.18 (0.20)
LSMO-DP1	834.2	837.8	1.05 (1.00)	0.48 (0.50)	641.8	642.8	1.32	0.18 (0.20)
LSMO-DP3	834.3	837.9	1.04 (1.00)	0.50 (0.50)	641.8	642.7	1.45	0.19 (0.20)
LSMO-DP5	834.4	837.6	1.02 (1.00)	0.51 (0.50)	641.7	642.6	1.20	0.19 (0.20)
LSMO-LP1	834.5	837.7	0.98 (1.00)	0.48 (0.50)	641.6	642.7	1.21	0.18 (0.20)
LSMO-PE1	834.4	837.8	0.92 (1.00)	0.49 (0.50)	641.9	642.8	1.16	0.18 (0.20)
LSMO-DP3-850	834.5	837.8	0.94 (1.00)	0.49 (0.50)	641.8	642.9	1.01	0.18 (0.20)
LSMO-DP3-900	834.3	837.9	0.95 (1.00)	0.51 (0.50)	641.9	642.7	0.93	0.17 (0.20)

Catalyst	O				H_2 consumption[e] (mmol/g)		Methane combustion (°C)		
	H_2O	O^{c}_{ads}	O^{d}_{lat}	O_{ads}/O^{2-}_{lat}	50–550 °C	550–950 °C	$T_{10\%}$	$T_{50\%}$	$T_{90\%}$
	BE	BE	BE	Molar ratio					
1D LSMO	532.8	530.9	529.3	0.81	1.43	1.52	562	672	749
LSMO-DP1	532.7	530.7	529.2	1.01	1.84	1.45	443	574	672
LSMO-DP3	532.9	530.8	529.3	1.22	1.86	1.49	437	566	661
LSMO-DP5	532.8	530.9	529.2	0.99	1.71	1.75	457	579	674
LSMO-LP1	532.8	530.8	529.3	0.98	1.66	1.78	483	601	687
LSMO-PE1	532.6	530.8	529.1	0.98	1.90	2.10	496	612	698
LSMO-DP3-850	532.8	530.9	529.2	0.97	1.54	1.65	482	566	701
LSMO-DP3-900	532.7	530.7	529.3	0.96	1.51	1.61	513	625	706

[a]Binding energy (eV)
[b]These data in parenthesis are the nominal La/Mn and/or Mn/(La + Mn + O) atomic ratios
[c]The adsorbed oxygen species
[d]The lattice oxygen species
[e]Data based on quantitative analysis of H_2-TPR profiles

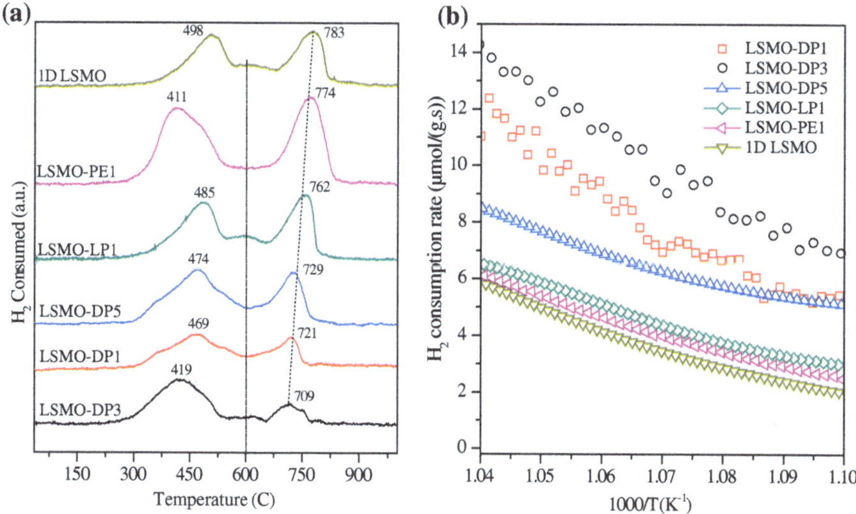

Fig. 4.7 **a** H_2-TPR profiles and **b** initial H_2 consumption rate as a function of inverse temperature of the LSMO catalysts

reduction peaks, the H_2 consumption in the low-temperature region (<500 °C) was 1.43, 1.51, 1.54, 1.90, 1.66, 1.71, 1.84, and 1.86 mmol/g, while that in the high-temperature zone (>500 °C) was 1.52, 1.61, 1.65, 2.10, 1.78, 1.75, 1.49, and 1.45 mmol/g for the 1D LSMO, LSMO-DP3-900, LSMO-DP3-850, LSMO-PE1, LSMO-LP1, LSMO-DP5, LSMO-DP3, and LSMO-DP1 catalysts, respectively. The H_2-TPR profiles of the catalysts with lower BET were similar to that of the 1D LSMO catalyst, while the peak at high temperatures (709–783 °C) gradually moved to higher temperatures and raised in intensity with the increase in BET. Meanwhile, each of the 3DOM material catalysts has shown a lower onset reduction temperature than the 1D catalyst. Moreover, from Table 4.2, the low-temperature H_2 consumptions of the LSMO catalysts were larger than that of the 1D sample counterpart. Figure 4.7b illustrates the initial H_2 consumption rate versus converse temperature. One can find that the initial H_2 consumption rate pursued a series of LSMO-DP3 > LSMO-DP1 > LSMO-DP5 > LSMO-LP1 > LSMO-PE1 > LSMO-DP3-850 > LSMO-DP3-900 > 1D LSMO, in close agreement with the sequence in BET of the LSMO structures.

4.3 Catalytic Performance

CH_4 conversion and CH_4 reaction rate rose with the increase in temperature, and the porous LSMO-DP3 sample achieved much better than the 1D LSMO sample (Fig. 4.8a and Table 4.2). It should be mentioned that methane oxidation in an empty microreactor (only quartz sands were loaded) was insignificant below ca.

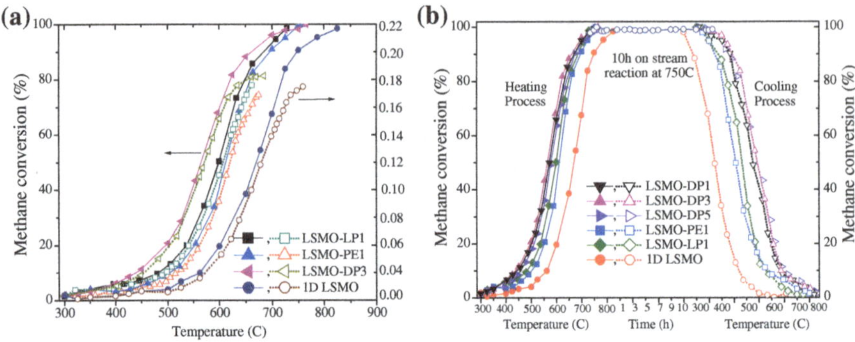

Fig. 4.8 **a** Methane conversion and methane reaction rate versus temperature over the LSMO-LP1, LSMO-PE1, LSMO-DP3, and 1D LSMO catalysts, **b** methane conversion versus temperature over the LSMO-DP1, LSMO-DP3, LSMO-DP5, LSMO-PE1, LSMO-LP1, and 1D LSMO catalysts

776 °C. That is to say, there was no homogeneous reaction when the temperature was lower than ca. 776 °C.

4.3.1 Study on Different Surfactant Added to the Catalyst

It is suitable to evaluate the catalytic performance of the catalysts by means of the reaction temperatures $T_{10\%}$, $T_{50\%}$, and $T_{90\%}$ as summarized in Table 4.2. It can be noted that the catalytic activities reduced in the following order: LSMO-DP3 > LSMO-DP1 > LSMO-DP5 > LSMO-LP1 > LSMO-PE1 > LSMO-DP3-850 > LSMO-DP3-900 > 1D LSMO, coinciding with the order in low-temperature reducibility (Fig. 4.8b). Clearly, the LSMO-DP3 sample with a BET of 42.1 m^2/g performed the high activity, giving the $T_{10\%}$, $T_{50\%}$, and $T_{90\%}$ of 437, 566, and 661 °C, respectively, which were much lower by 125, 106, and 88 °C than those ($T_{10\%}$ = 562 °C, $T_{50\%}$ = 672 °C, and $T_{90\%}$ = 749 °C) achieved over the 1D LSMO catalyst with a BET of 2.6 m^2/g, respectively. Additionally, onstream reaction of the LSMO-DP3 catalysts was tested within 10 h at a GHSV of 30,000 ml/(g h) and a temperature of 750 °C, as illustrated in Fig. 4.8b. Clearly, no significant loss in catalytic performance was detected. In addition, performance of the 3DOM LSMO catalysts followed the same order in the cases of heating and cooling runs.

4.3.2 Influence of the Temperature of Calcination on the Catalyst

With the increase in calcination temperature, the obtained LSMO-DP3-850 and LSMO-DP3-900 samples illustrated performance lower than that ($T_{50\%}$ = 566 °C

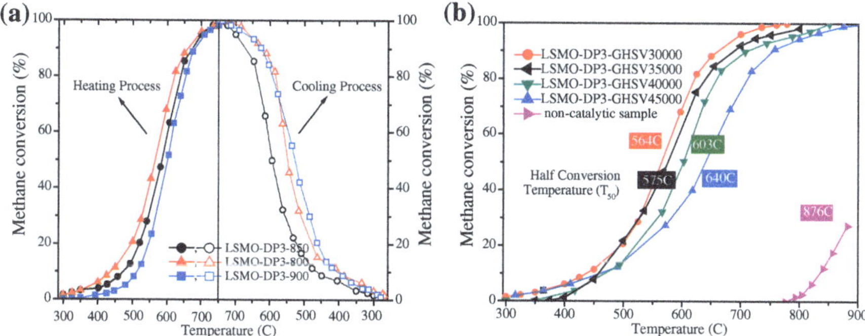

Fig. 4.9 **a** Methane conversion versus temperature over the LSMO-DP3-800, LSMO-DP3-850, and LSMO-DP3-900 catalysts, and **b** effect of GHSV on the catalytic activity over the LSMO-DP3 catalyst

and $T_{90\%}$ = 661 °C) of the LSMO-DP3 sample obtained after calcination at 800 °C (Fig. 4.9a). Meanwhile, Fig. 4.9b illustrates the catalytic performance of the LSMO-DP3 catalyst at various GHSVs. At the same temperature, CH_4 conversion dropped with the increase in GHSV. This result suggests that a longer contact time was helpful for the improvement in catalytic performance. In order to study the stability of performance of catalysts, we carried out the lifetime experiment over the LSMO-DP3 sample within 24 h of onstream reaction (Fig. A.10 of the Appendix) and recorded the XRD pattern and HRSEM image of the used LSMO-DP3 sample.

Turnover frequencies (TOFs) of the 3DOM LSMO and 1D LSMO samples for CH_4 combustion were summarized in Table 4.3. It is noted that the TOF_{LSMO} value of each of the LSMO samples rose with the increase in temperature. In the same temperature, the 3DOM LSMO catalysts showed much higher TOF_{LSMO} values than the 1D LSMO catalysts. With the increase in DMOTEG concentration from 1.00 to 3.00 ml during the synthesis procedure, the obtained LSMO-DP1 and LSMO-DP3 catalysts show the TOF_{LSMO} values of 6.80×10^{-6} and 8.17×10^{-6} $mol_{methane}/(mol_{LSMO}\ s)$ at 300 °C, and 5.14×10^{-5} and 7.92×10^{-5} $mol_{methane}/(mol_{LSMO}\ s)$ at 400 °C, respectively, while dropped to 5.77×10^{-6} $mol_{methane}/(mol_{LSMOs})$ at 300 °C and 4.49×10^{-5} $mol_{methane}/(mol_{LSMOs})$ at 400 °C with a further increase in DMOTEG concentration from 3.00 to 5.00 ml.

It is worth noting out that under similar reaction conditions, the performance ($T_{50\%}$ = 566 °C and $T_{90\%}$ = 661 °C) over 3DOM LSMO-DP3 sample was much better than those ($T_{50\%}$ = 662 °C and $T_{90\%}$ = 739 °C) over La_2CuO_4 nanorods [25] ($T_{50\%}$ = 654 °C and $T_{90\%}$ = 800 °C) over $La_{0.9}Cu_{0.1}MnO_3$ [26] ($T_{50\%}$ = 620 °C and $T_{90\%}$ = 710 °C) over $La_{0.5}Sr_{0.5}MnO_3$ [27] ($T_{50\%}$ = 710 °C and $T_{90\%}$ = 770 °C) over 20 wt% $LaMnO_3/MgO$ [28], whereas inferior to those ($T_{50\%}$ = 520 °C and $T_{90\%}$ = 570 °C) over $La_{0.5}Sr_{0.5}MnO_3$ cubes [29] and those ($T_{50\%}$ = 365 °C and $T_{90\%}$ = 425 °C) over 1 wt% Pd/ZrO_2 [30].

Table 4.3 Turnover frequencies (TOF_{LSMO}), rate constants (k), reaction rates (r), pre-exponential factors (A), activation energies (E_a), and correlation coefficients (R^2) of the 3DOM LSMO and 1D LSMO catalysts for methane combustion in the range of 300–550 °C

Catalyst	TOF_{LSMO} ($\mu mol_{methane}/(mol_{LSMO}$ s))				k ($\times 10^{-2}$ s^{-1})/r ($\times 10^{-3}$ μmol/(g s))[a]				Kinetic parameter		
	300 °C	325 °C	350 °C	400 °C	300 °C	400 °C	500 °C	550 °C	A (s^{-1})	E_a (kJ/mol)	R^2
1D LSMO	0.057	0.84	1.25	1.73	1.67/3.74	2.51/5.56	3.50/7.69	4.26/9.28	9.43×10^3	91.4	0.9991
LSMO-DP1	0.680	1.54	3.47	5.14	9.52/19.7	12.7/25.6	20.1/38.1	31.5/54.4	3.54×10^3	65.1	0.9995
LSMO-DP3	0.817	2.22	5.22	7.92	6.73/14.3	9.54/19.7	13.1/26.3	26.4/47.4	3.67×10^3	56.6	0.9990
LSMO-DP5	0.577	1.08	3.32	4.49	6.95/14.7	9.64/19.9	15.4/30.4	22.3/41.5	2.55×10^3	67.5	0.9982
LSMO-LP1	0.791	1.56	2.96	3.62	7.64/16.1	10.6/21.7	15.2/29.9	25.0/45.4	3.76×10^3	73.1	0.9997
LSMO-PE1	0.970	1.48	2.14	3.21	4.38/9.53	6.72/14.3	9.64/19.9	14.4/28.6	4.09×10^3	75.2	0.9991
LSMO-DP3-850	0.174	1.35	3.40	5.32	3.24/7.65	4.59/8.92	7.91/6.37	9.43/12.1	7.13×10^4	66.4	0.9993
LSMO-DP3-900	0.091	1.07	2.96	4.19	2.74/5.88	3.74/4.42	5.21/6.85	8.49/9.32	8.54×10^5	81.2	0.9992

[a]The data before the "/" symbol are the k values, whereas those after the "/" symbol are the r values

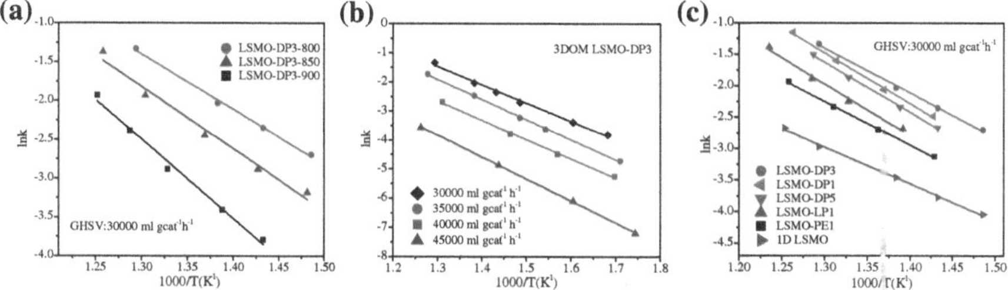

Fig. 4.10 Arrhenius plots for methane combustion on the as-prepared porous LSMO catalysts

4.3.2.1 Apparent Activation Energy

Figure 4.10 illustrates the Arrhenius plots for CH_4 combustion at methane conversion <25 % (at which the temperature range was 300–550 °C) over the LSMO samples. Based on the slopes of the well-linear (correlation coefficients (R^2) were above 0.99) Arrhenius plots, one can determine the E_a, A, and k values for methane combustion over these samples, as can be seen in Table 4.3. It can be obviously seen that the E_a value (91.4 kJ/mol) of the 1D LSMO sample was much higher than those (56.5–75.2 kJ/mol) of the porous LSMO samples, although the 3DOM LSMO-DP3 as well as LSMO-DP1 samples displayed the lowest E_a values (56.6–65.1 kJ/mol). The A value dropped based on the order of LSMO-DP3 > LSMO-DP1 > LSMO-DP5 > LSMO-LP1 > LSMO-PE1 > LSMO-DP3-850 > LSMO-DP3-900 > 1D LSMO. By comparing the E_a values of different samples, one can estimate their catalytic activities. The lower the E_a value, the easier is the complete oxidation of CH_4 over LSMO samples and, for this reason, higher catalytic activity. Table 4.3 illustrated that the apparent activation energies (56.6–67.5 kJ/mol) obtained over 3DOM architecture LSMO-DP samples were close to or a little lower than those (62 kJ/mol) over $Pt/Ce_{0.64}Zr_{0.15}Bi_{0.21}O_{1.895}/Al_2O_3$ [31] and those (73–89 kJ/mol) over $M_xFe_{3-x}O_4$ (M = Ni, Mn; x = 0.5–0.65) [32] but much lower than those (120–144 kJ/mol) over CuO/Al_2O_3 and MnO/Al_2O_3 [33], similar to those (51–79 kJ/mol) over 10–20 wt% $LaCoO_3/Ce_{1-x}Zr_xO_2$ (x = 0–0.2) [34]. As a result, the findings of kinetic studies approve that the 3DOM LSMO samples demonstrated excellent catalytic activity for the CH_4 combustion, and it is a promising catalyst for the combustion of CH_4 in practical application.

4.4 Conclusion and Discussion

1. We successfully synthesized rhombohedrally crystallized, three-dimensionally ordered macroporous $La_{0.6}Sr_{0.4}MnO_3$ (3DOM LSMO) by the surfactant-mediated colloidal crystal PMMA-templating route. The macropores are highly ordered and interconnected each other by tiny pore windows. The LSMO-DP3

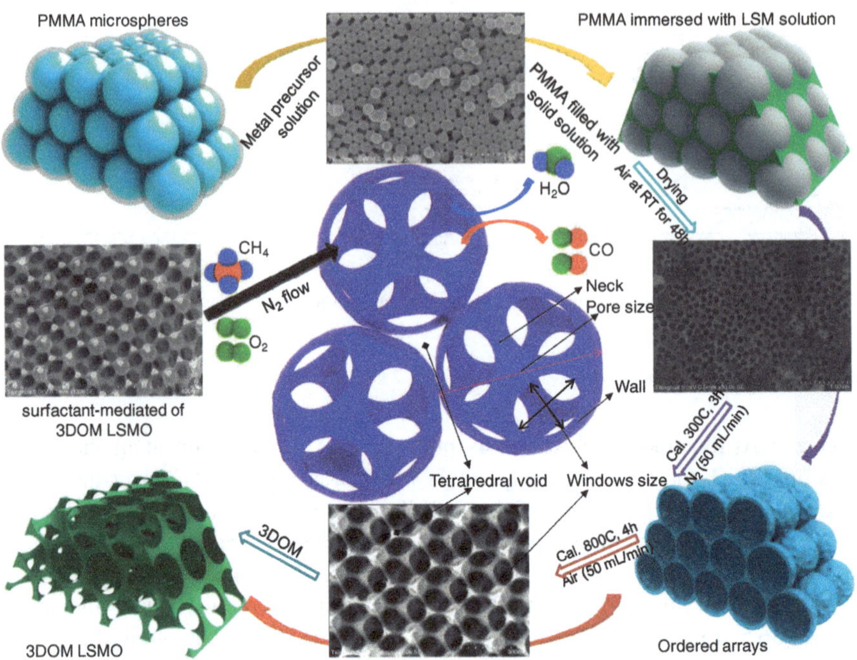

Scheme 4.1 Rhombohedrally crystallized 3DOM $La_{0.6}Sr_{0.4}MnO_3$ (LSMO) catalysts with a high surface area of 32–42 m^2/g are prepared by the PMMA-templating method. The 3DOM LSMO-DP3 catalyst derived with 3.0 ml of dimethoxytetraethylene glycol and 5.0 ml of polyethylene glycol shows the excellent activity for the combustion of methane

catalyst with a surface area 42.1 m^2/g performed the best, success the $T_{10\%}$, $T_{50\%}$, and $T_{90\%}$ of 437, 566, and 661 °C, respectively. They were much lower by 125, 106, and 88 °C than those ($T_{10\%}$ = 562 °C, $T_{50\%}$ = 672 °C, and $T_{90\%}$ = 749 °C) obtained on the 1D LSMO catalyst with a BET of 2.6 m^2/g, respectively. These results propose that the useful contact between catalyst and feed is one of the most significant factors to influence the performance of CH_4 combustion.

2. The thermal stability of LSMO catalyst was very good. After being aged at 800 °C, the 3DOM structure was still maintained perfectly. In addition, the DMOTEG/PEG400 volumetric ratio had an influence on the crystalline phase, crystallite size, and crystallinity of 3DOM LSMO. There was the copresence of surface Mn^{4+} and Mn^{3+} species on LSMO, and the surface Mn^{4+}/Mn^{3+} molar ratios of LSMO-DP3 and LSMO-DP1 were much higher than those of the other LSMO samples.
3. It was shown that compared to the 1D LSMO, the 3DOM LSMO-DP3 sample shows a much higher TOF_{LSMO} value (6.64 × 10^{-8} and 1.99 × 10^{-7} $\mu mol_{methane}/(cm^2\ s)$ at 300 and 400 °C, respectively). The apparent activation energies of the 3DOM LSMO catalysts were estimated to be 56.5–75.2 kJ/mol, with the LSMO-DP3 catalyst exhibiting the lowest apparent activation energy

(56.6 kJ/mol). It is concluded that the good catalytic activity of LSMO-DP3 for CH_4 combustion was associated with its larger BET, higher O_{ads} species concentration, and better low-temperature reducibility as well as unique nanovoid-containing 3DOM architecture (Scheme 4.1).

References

1. Auer R, Alifanti M, Delmon B, Thyrion FC. Catalytic combustion of methane in the presence of organic and inorganic compounds over $La_{0.9}Ce_{0.1}CoO_3$ catalyst. Appl Catal B. 2002;39 (4):311–8.
2. Zhong Z, Chen K, Ji Y, Yan Q. Methane combustion over B-site partially substituted perovskite-type $LaFeO_3$ prepared by sol–gel method. Appl Catal A. 1997;156(1):29–41.
3. Machida M, Eguchi K, Arai H. Effect of structural modification on the catalytic property of Mn-substituted hexaaluminates. J Catal. 1990;123(2):477–85.
4. Duprat AM, Alphonse P, Sarda C, Rousset A, Gillot B. Nonstoichiometry-activity relationship in perovskite-like manganites. Mater Chem Phys. 1994;37(1):76–81.
5. Niu J, Deng J, Liu W, et al. Nanosized perovskite-type oxides $La_{1-x}Sr_xMO_{3-\delta}$ for the catalytic removal of ethylacetate. Catal Today. 2007;126(3):420–9.
6. Zhang G, Wang D, Möhwald H. Patterning microsphere surfaces by templating colloidal crystals. Nano Lett. 2004;5(1):143–6.
7. Kakihana M. Invited review "sol–gel" preparation of high temperature superconducting oxides. J Sol–Gel Sci Technol. 1996;6(1):7–55.
8. Rivas I, Alvarez J, Pietri E, Pérez-Zurita MJ, Goldwasser MR. Perovskite-type oxides in methane dry reforming: Effect of their incorporation into a mesoporous SBA-15 silica-host. Catal Today. 2010;149(3):388–93.
9. Liang JJ, Weng H-S. Catalytic properties of lanthanum strontium transition metal oxides ($La_{1-x}Sr_xBO_3$; B = manganese, iron, cobalt, nickel) for toluene oxidation. Ind Eng Chem Res. 1993;32(11):2563–72.
10. Kakihana M. Invited review "sol–gel" preparation of high temperature superconducting oxides. J Sol–Gel Sci Technol. 1996;6(1):7–55.
11. Zheng J, Liu J, Zhao Z, Xu J, Duan A, Jiang G. The synthesis and catalytic performances of three-dimensionally ordered macroporous perovskite-type $LaMn_{1-x}Fe_xO_3$ complex oxide catalysts with different pore diameters for diesel soot combustion. Catal Today. 2012;191 (1):146–53.
12. Sadakane M, Kato R, Murayama T, Ueda W. Preparation and formation mechanism of three-dimensionally ordered macroporous (3DOM) MgO, $MgSO_4$, $CaCO_3$, and $SrCO_3$, and photonic stop band properties of 3DOM $CaCO_3$. J Solid State Chem. 2011;184(8):2299–305.
13. Sadakane M, Sasaki K, Kunioku H, Ohtani B, Ueda W, Abe R. Preparation of nano-structured crystalline tungsten (vi) oxide and enhanced photocatalytic activity for decomposition of organic compounds under visible light irradiation. Chem Commun. 2008;1(48):6552–4.
14. Sadakane M, Sasaki K, Nakamura H, Yamamoto T, Ninomiya W, Ueda W. Important property of polymer spheres for the preparation of three-dimensionally ordered macroporous (3DOM) metal oxides by the ethylene glycol method: the glass-transition temperature. Langmuir. 2012;28(51):17766–70.
15. Ji K, Dai H, Deng J, et al. Three-dimensionally ordered macroporous $Eu_{0.6}Sr_{0.4}FeO_3$ supported cobalt oxides: highly active nanocatalysts for the combustion of toluene. Appl Catal B. 2013;129(1):539–48.

16. Liu Y, Dai H, Deng J, Zhang L, Au CT. Three-dimensional ordered macroporous bismuth vanadates: PMMA-templating fabrication and excellent visible light-driven photocatalytic performance for phenol degradation. Nanoscale. 2012;4(7):2317–25.
17. Li H, Zhang L, Dai H, He H. Facile synthesis and unique physicochemical properties of three-dimensionally ordered macroporous magnesium oxide, gamma-alumina, and ceria–zirconia solid solutions with crystalline mesoporous walls. Inorg Chem. 2009;48(10):4421–34.
18. Hornés A, Gamarra D, Munuera G, Conesa JC, Martínez-Arias A. Catalytic properties of monometallic copper and bimetallic copper-nickel systems combined with ceria and Ce-X mixed oxides applicable as SOFC anodes for direct oxidation of methane. J Power Sources. 2007;169(1):9–16.
19. Hornés A, Gamarra D, Munuera G, et al. Structural, catalytic/redox and electrical characterization of systems combining Cu–Ni with CeO_2 or $Ce_{1-x}M_xO_{2-\delta}$ (M = Gd or Tb) for direct methane oxidation. J Power Sources. 2009;192(1):70–7.
20. Ruckenstein E, Hu YH. Carbon dioxide reforming of methane over nickel/alkaline earth metal oxide catalysts. Appl Catal A. 1995;133(1):149–61.
21. Deng J, Zhang L, Dai H, He H, Au CT. Hydrothermally fabricated single-crystalline strontium-substituted lanthanum manganite microcubes for the catalytic combustion of toluene. J Mol Catal A: Chem. 2009;299(1):60–7.
22. Di Castro V, Polzonetti G. XPS study of MnO oxidation. J Electron Spectrosc Relat Phenom. 1989;48(1):117–23.
23. Liu Y, Dai H, Du Y, Deng J, Zhang L, Zhao Z. Lysine-aided PMMA-templating preparation and high performance of three-dimensionally ordered macroporous $LaMnO_3$ with mesoporous walls for the catalytic combustion of toluene. Appl Catal B. 2012;119(1):20–31.
24. Fierro JLG, Tascón JMD, Tejuca LG. Physicochemical properties of $LaMnO_3$: reducibility and kinetics of O_2 adsorption. J Catal. 1984;89(2):209–16.
25. Zhang L, Zhang Y, Dai H, Deng J, Wei L, He H. Hydrothermal synthesis and catalytic performance of single-crystalline $La_{2-x}Sr_xCuO_4$ for methane oxidation. Catal Today. 2010;153(3):143–9.
26. Bulgan G, Teng F, Liang SH, Yao WQ, Zhu YF. Effect of Cu doping on the structure and catalytic activity of $LaMnO_3$ catalyst. Wuli Huaxue Xuebao/Acta Phys Chim Sin. 2007;23 (9):1387–92.
27. de Araujo GC, Lima S, Rangel MdC, Parola VL, Peña MA, García Fierro JL. Characterization of precursors and reactivity of $LaNi_{1-x}Co_xO_3$ for the partial oxidation of methane. Catal Today. 2005;107(1):906–12.
28. Svensson EE, Nassos S, Boutonnet M, Järås SG. Microemulsion synthesis of MgO-supported $LaMnO_3$ for catalytic combustion of methane. Catal Today. 2006;117(4):484–90.
29. Araya P, Guerrero S, Robertson J, Gracia FJ. Methane combustion over Pd/SiO_2 catalysts with different degrees of hydrophobicity. Appl Catal A. 2005;283(1):225–33.
30. Liang S, Teng F, Bulgan G, Zhu Y. Effect of Jahn—Teller distortion in $la_{0.5}Sr_{0.5}MnO_3$ cubes and nanoparticles on the catalytic oxidation of CO and CH_4. J Phys Chem C. 2007;111 (45):16742–9.
31. Florea M, Alifanti M, Parvulescu VI, et al. Total oxidation of toluene on ferrite-type catalysts. Catal Today. 2009;141(3):361–6.
32. Masui T, Imadzu H, Matsuyama N, Imanaka N. Total oxidation of toluene on $Pt/CeO_2–ZrO_2–Bi_2O_3/\gamma\text{-}Al_2O_3$ catalysts prepared in the presence of polyvinyl pyrrolidone. J Hazard Mater. 2010;176(1):1106–9.
33. Saqer SM, Kondarides DI, Verykios XE. Catalytic oxidation of toluene over binary mixtures of copper, manganese and cerium oxides supported on $\gamma\text{-}Al_2O_3$. Appl Catal B. 2011;103 (3):275–86.
34. Florea M, Alifanti M, Parvulescu VI, et al. Total oxidation of toluene on ferrite-type catalysts. Catal Today. 2009;141(3):361–6.

Chapter 5
3DOM LSMO-Supported Ag NPs for Catalytic Combustion of Methane

5.1 Introduction

Perovskite-type oxides (ABO_3) have attracted a lot of attention in the last few decades [1–7]. The major drawback of traditional ABO_3 is low in surface area, which limits its catalytic application. In fact, preparation of ABO_3 with sufficiently high surface areas is a big challenge since it involves a solid-state reaction of the precursors to form the ABO_3 structure [7]. The colloidal crystal template can be fabricated by ordering monodispersed microspheres (e.g., polystyrene (PS), polymethyl methacrylate (PMMA), or silica) into a face-centered close-packed array [8–12]. Previously, we found that when La^{3+} was partially substituted by Sr^{2+} in $LaMnO_3$ could lead to increases in Mn^{4+} concentration and oxygen vacancy density, and thus enhancing the catalytic performance. When the x value in $La_{1-x}Sr_xFeO_3$ was equal to 0.4, the performance of catalyst was best for the combustion of volatile organic compounds [13–16]. It has been demonstrated that doping of perovskites such as $LaNiO_3$, $LaMnO_3$ [17, 18], $La_{0.5}Sr_{0.5}CoO_3$ [19], and $LaFe_{0.5}Co_{0.5}O_3$ [20], $LaFeO_3$, and Ag increases their activities in methane combustion. Gardner et al. [21] studied the manganese oxide catalysts with low Ag loadings (≤1 %) for low-temperature CO oxidation and observed good catalytic activities. Moreover, Ag-doped $LaCoO_3$ showed no deactivation in methane combustion during 50 h of onstream reaction at 600 °C [17]. Even greater activity enhancement in comparison with $LaMnO_3$ was observed in the case of Ag/MnO_x/perovskite composite catalysts in CO oxidation [22]. The ABO_3 support used in most of the studies, however, are all nonporous. Since porous ABO_3 can provide good dispersion of active components on the surface, it would be better to prepare the Ag/3DOM-ABO_3 catalysts that show large surface areas and porous structures. In the past several years, other groups have extended attention to the synthesis and physicochemical property characterization of 3DOM-structured materials (e.g., Co_3O_4/3DOM $La_{0.6}Sr_{0.4}CoO_3$ (BET = 29–32 m^2/g) [23], yCrOx/3DOM $InVO_4$ (BET = 41.3–52.3 m^2/g) [24], 3DOM $InVO_4$ (BET = 35–52 m^2/g) [25],

H. Arandiyan, *Methane Combustion over Lanthanum-based Perovskite Mixed Oxides*, Springer Theses, DOI 10.1007/978-3-662-46991-0_5

*x*Au/3DOM $LaMnO_3$ (BET = 29.8–32.7 m^2/g) [26], Au/3DOM $LaCoO_3$ (BET = 24–29 m^2/g) [27], and Au/3DOM $La_{0.6}Sr_{0.4}MnO_3$ (BET = 31.1–32.9 m^2/g) [28] by the surfactant-assisted PMMA-templating approaches and observed that some of the 3DOM materials showed excellent catalytic performance in the combustion of toluene and methane [29–31]. Herein, we report the followings for the first time: the preparation, characterization, and catalytic methane combustion activities of 3DOM $La_{0.6}Sr_{0.4}MnO_3$ with nanovoid-like or mesoporous skeletons and its supported Ag NPs nanocatalyst (*y*Ag/3DOM $La_{0.6}Sr_{0.4}MnO_3$; $y = 0$, 1.57, 3.63, and 5.71 wt%) via the dimethoxytetraethylene glycol (DMOTEG)-assisted gas bubbling reduction route with well-arrayed PMMA microspheres as hard template. It was found that the DMOTEG-mediated reduction strategy not only produced size-controlled Ag NPs, but also stabilized them against conglomeration without the need of additional stabilizers. By using this novel method, Ag NPs could be highly dispersed on 3DOM $La_{0.6}Sr_{0.4}MnO_3$, and the obtained catalysts showed excellent performance and stability for methane combustion.

5.2 Characterization and Activity Evaluation of *y*wt% Ag/3DOM LSMO

5.2.1 X-ray Diffraction (XRD) Patterns

Figure 5.1 shows the XRD patterns of 1D $La_{0.6}Sr_{0.4}MnO_3$, 3DOM $La_{0.6}Sr_{0.4}MnO_3$, and *y*Ag/3DOM $La_{0.6}Sr_{0.4}MnO_3$ ($y = 1.57$, 3.63, and 5.71 wt%) catalysts. By comparing the XRD pattern of the standard $La_{0.6}Sr_{0.4}MnO_3$ sample (JCPDS PDF# 50-0308), one can deduce that all of the Bragg diffraction peaks in the 2θ range of 8–90° could be well indexed, as indicated in Fig. 5.1. The characteristic diffraction

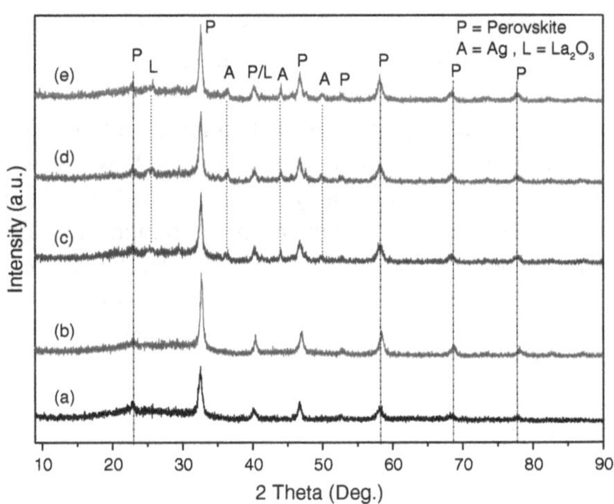

Fig. 5.1 XRD patterns of (*a*) 1D $La_{0.6}Sr_{0.4}MnO_3$, (*b*) 3DOM $La_{0.6}Sr_{0.4}MnO_3$, (*c*) 1.58 wt% Ag/3DOM $La_{0.6}Sr_{0.4}MnO_3$, (*d*) 3.63 wt% Ag/3DOM $La_{0.6}Sr_{0.4}MnO_3$, (*e*) 5.71 wt% Ag/3DOM $La_{0.6}Sr_{0.4}MnO_3$

peaks with 2θ values of 22.9°, 32.6°, 40.2°, 46.8°, 52.9°, 58.1°, 68.3°, and 77.9° are corresponded to the (012), (110), (202), (024), (116), (300), (208), and (128) lattice planes of the rhombohedral perovskite phase [31], respectively. The Bragg diffraction intensity of the 3DOM-structured samples was higher than that of the 1D counterpart, indicating that the former possessed better crystallinity. The crystal sizes of *y*Ag/3DOM $La_{0.6}Sr_{0.4}MnO_3$ were basically the same (23.9–24.6 nm) and much lower than that (61.4 nm) of the 1D $La_{0.6}Sr_{0.4}MnO_3$ sample, which are calculated by Scherrer equation using the FWHM of the (110) crystal face line of $La_{0.6}Sr_{0.4}MnO_3$, and the calculated values are summarized in Table 5.1. When the loading of Ag NPs increased to 5.71 wt%, the crystal size of 5.71 wt% Ag/3DOM $La_{0.6}Sr_{0.4}MnO_3$ descended to ca. 23.9 nm.

It is clearly seen from the XRD patterns (Fig. 5.1c–e) that three weak peaks at $2\theta = 36.4°$, 44.1°, and 49.8° were observed which became more pronounced as the Ag loading increased from 1.58 to 5.71 wt%. This is in agreement with the known 2θ values for metallic silver [32]. Regrettably, it has been impossible to ascertain whether or not Ag_2O is present in the perovskite samples because the main Ag_2O (111) peak at $2\theta = 32.5°$ and the $LaMnO_3$ perovskite peak overlap [31, 32]. Therefore, there is no conclusive evidence for the existence of Ag_2O crystal phase on the *y*Ag/3DOM $La_{0.6}Sr_{0.4}MnO_3$ samples. These results are in good agreement with those of $La_{1-x}Ag_xMnO_3$ (at $x = 0.2$ [33] and 0.22 [34] or $x > 0.25$ [35]) that contained a rhombohedral perovskite and a metal Ag phase.

5.2.2 Inductively Coupled Plasma-Atomic Emission Spectroscopy (ICP-AES)

In the meantime, the elemental analyses of the actual Ag contents in the *y*Ag/3DOM $La_{0.6}Sr_{0.4}MnO_3$ samples were determined by the ICP-AES technique, and the results are summarized in Table 5.1. It can be seen that the actual Ag content in each sample was lower than the initial theoretical value during the preparation process, indicating that most of the Ag in the aqueous solution was deposited on the surface of 3DOM $La_{0.6}Sr_{0.4}MnO_3$.

5.2.3 Thermo Gravimetric Analysis (TGA) and (FT-IR) Spectroscopy

Furthermore, the calcination conditions were appropriate for the generation of single-phase perovskite structure and total removal of the organics (DMOTEG and PEG400) as well as the hard template (PMMA), as substantiated by the TGA/DTA and FT-IR results of the typical samples (Fig. 5.2a and b). It should be noted that no detection of sulfate groups in the FT-IR spectra of the typical PMMA (Fig. 5.2b)

Table 5.1 Preparation conditions, theoretical Ag loading (T.V.), real Ag content (R.V.), Ag particle sizes, crystallite sizes, BET surface areas, pore diameters, and pore volumes of the 1D $La_{0.6}Sr_{0.4}MnO_3$, 3DOM $La_{0.6}Sr_{0.4}MnO_3$, and *y*Ag/3DOM $La_{0.6}Sr_{0.4}MnO_3$ samples

Catalyst	Method	Hard template/soft template	Calcination condition	Ag loading (wt%)[a]		Ag particle size (nm)[b]	XRD result		BET surface area (m^2/g)			Pore diameter (nm)		Pore volume (cm^3/g)
				T.V.	R.V.		Crystal phase	Crystallite size[c] (nm)	Macropore (≥50 nm)	Mesopore (<50 nm)	Total	Macropore (≥50 nm)[d]	Mesopore (<50 nm)[e]	
1D $La_{0.6}Sr_{0.4}MnO_3$	Citric acid	–/–	850 °C, air (50 ml/min), 4 h	–/–	–/–	–/–	Rhombohedral	61.4	–/–	–/–	2.6	–/–	–/–	–/–
3DOM $La_{0.6}Sr_{0.4}MnO_3$	PMMA-templating	PMMA/(DMOTEG-PEG400)	300 °C, N_2 (50 ml/min), 3 h → 300 °C, air (50 ml/min), 1 h → 800 °C, air (50 ml/min), 4 h	–/–	–/–	–/–	Rhombohedral	23.5	7.3	34.8	42.1	119	2.9–4.1	0.159
1.58 wt% Ag/3DOM $La_{0.6}Sr_{0.4}MnO_3$	In situ PMMA-templating	PMMA/(DMOTEG-PEG400)	300 °C, N_2 (50 ml/min), 3 h → 300 °C, air (50 ml/min), 1 h → 800 °C, air (50 ml/min), 4 h	2.0	1.58	3.3	Rhombohedral	24.2	7.1	35.6	42.7	105	3.4–5.6	0.161
3.63 wt% Ag/3DOM $La_{0.6}Sr_{0.4}MnO_3$	In situ PMMA-templating	PMMA/(DMOTEG-PEG400)	300 °C, N_2 (50 ml/min), 3 h → 300 °C, air (50 ml/min), 1 h → 800 °C, air (50 ml/min), 4 h	4.0	3.63	3.2	Rhombohedral	24.6	7.4	34.1	41.5	101	4.7–6.1	0.165
5.71 wt% Ag/3DOM $La_{0.6}Sr_{0.4}MnO_3$	In situ PMMA-templating	PMMA/(DMOTEG-PEG400)	300 °C, N_2 (50 ml/min), 3 h → 300 °C, air (50 ml/min), 1 h → 800 °C, air (50 ml/min), 4 h	6.0	5.71	3.5	Rhombohedral	23.9	7.6	30.6	38.2	109	5.6–6.9	0.144

[a]Determined by the ICP-AES technique
[b]Estimated according to the SEM and statistic analysis of more than 200 Ag particles in HRTEM images
[c]Data determined based on the XRD results according to the Scherrer equation using the FWHM of the (110) line of $La_{0.6}Sr_{0.4}MnO_3$
[d]Data were estimated according to the SEM and TEM images
[e]Data were estimated according to the pore-size distributions
[f]Data were estimated according to the BJH method

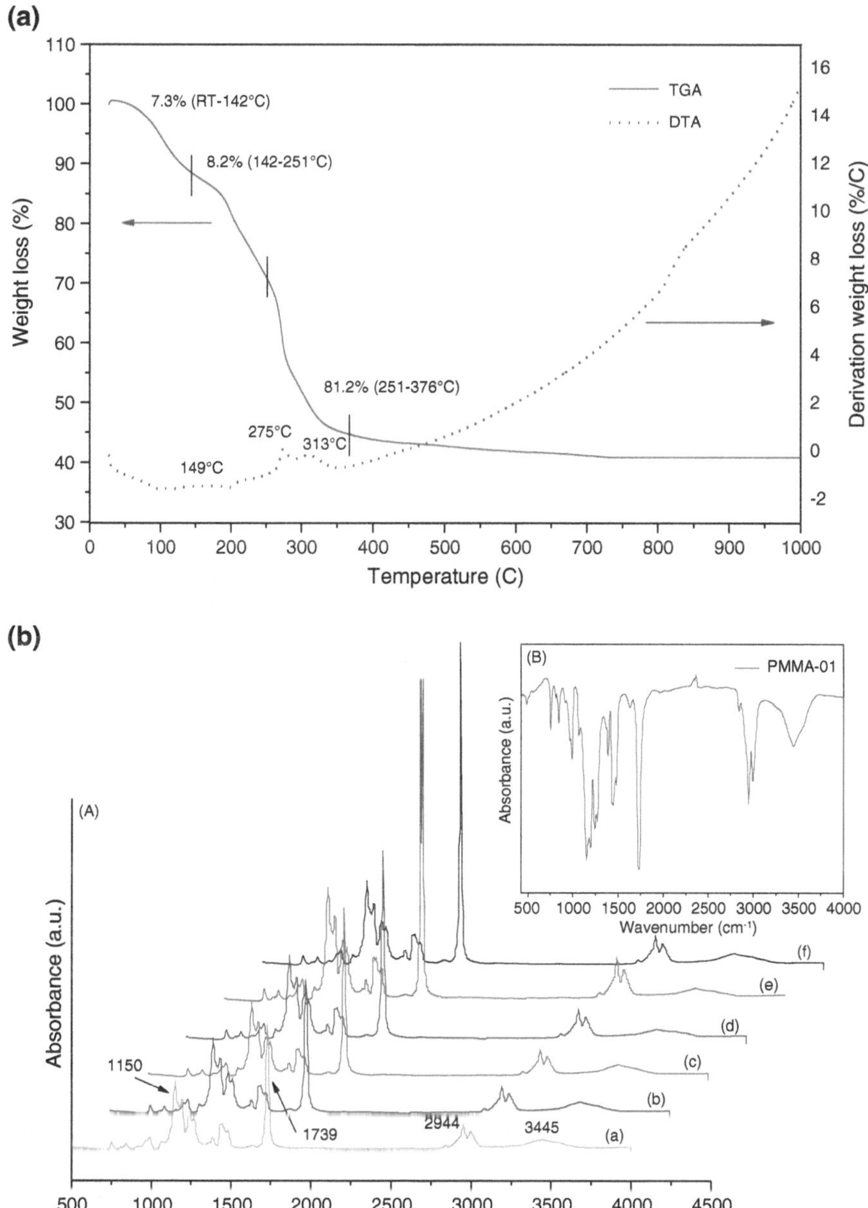

Fig. 5.2 **a** TGA/DTA profiles of the 3DOM $La_{0.6}Sr_{0.4}MnO_3$ sample before calcination at 300 °C in N_2 (50 ml/min) for 3 h; **b** the FT-IR spectra of *(a–f)* PMMA microspheres

and samples (Fig. A.1 of Appendix) before and after calcination in air at 800 °C for 4 h was indicative of the complete removal of sulfate groups after subsequent washing and filtration of the catalyst precursors. In the meantime, deposition of Ag NPs on the surface of 3DOM $La_{0.6}Sr_{0.4}MnO_3$ leads to less impact on the crystalline phase of 3DOM $La_{0.6}Sr_{0.4}MnO_3$.

5.2.4 HRSEM and EDS Results

Figure 5.3 shows the SEM images of the 1D $La_{0.6}Sr_{0.4}MnO_3$, 3DOM $La_{0.6}Sr_{0.4}MnO_3$, and *y*Ag/3DOM $La_{0.6}Sr_{0.4}MnO_3$ samples. It can be clearly observed that the macroporous *y*Ag/3DOM $La_{0.6}Sr_{0.4}MnO_3$ materials contain skeletons surrounding uniform close-packed periodic voids with an average diameter of 105 ± 20 nm, which corresponds to shrinkage of 20–35 % compared with the initial size (161 nm) of PMMA microspheres. A closer analysis in the HRSEM images of an ordered region reveals a structure with largely interconnected open windows (50 ± 15 nm in diameter) (Fig. 5.3). The *y*Ag/3DOM

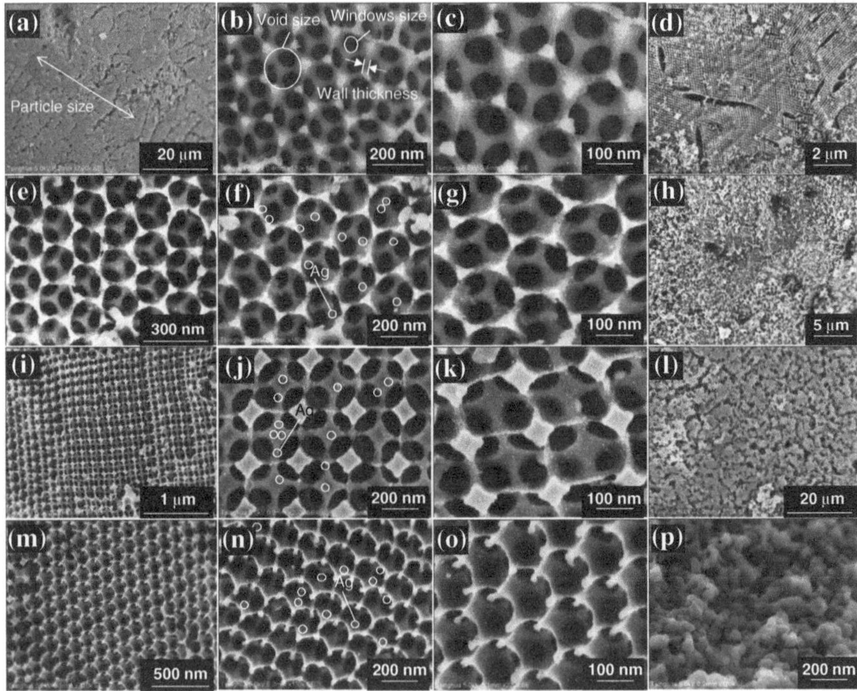

Fig. 5.3 SEM images of **a–c** 3DOM $La_{0.6}Sr_{0.4}MnO_3$, **d–g** 1.58 wt% Ag/3DOM $La_{0.6}Sr_{0.4}MnO_3$, **h–k** 3.63 wt% Ag/3DOM $La_{0.6}Sr_{0.4}MnO_3$, **l–o** 5.71 wt% Ag/3DOM $La_{0.6}Sr_{0.4}MnO_3$, and **p** 1D $La_{0.6}Sr_{0.4}MnO_3$

$La_{0.6}Sr_{0.4}MnO_3$ materials are similar to the well-defined 3DOM-architectured $La_{0.6}Sr_{0.4}MnO_3$ support. The introduced PEG400 reacted with the metal nitrates at a temperature more than 100 °C to form glyoxylate anion [36], which could function as a ligand to coordinate with the metal nitrates, thus generating the heterometallic complexes [37]. The glassy transition temperature of the PMMA template was ca. 370 °C in air [30, 31, 38], to which the decomposition temperature of manganese nitrate was close. During the calcination process in N_2 at 300 °C, the partial carbonization of PMMA took place, and the generated amorphous carbon would act as a hard template to guarantee the formation of high-quality 3DOM-structured $La_{0.6}Sr_{0.4}MnO_3$ before the complete oxidative removal of the PMMA template at 800 °C [30, 31]. These results indicate that the interconnected macroporous morphology was well formed after removing the template. With the loading of Ag NPs, the 3DOM structure was retained in the *y*Ag/3DOM $La_{0.6}Sr_{0.4}MnO_3$ samples. The introduction of an excessive amount of $Ag(NO_3)_2$ had no significant effects on the decrease in 3DOM quality (Fig. 5.3l–o). Meanwhile, the amount of DMOTEG (3.00 ml) in the presence of PEG400 (5.00 ml) added to the precursor solution during the preparation process exerted an impact on the crystalline size and led to the generation of a super high-quality 3DOM architecture with interconnected pore walls in the *y*Ag/3DOM $La_{0.6}Sr_{0.4}MnO_3$ samples (Fig. 5.3).

5.2.5 HRTEM and SAED Pattern Results

Figure 5.4 shows TEM and HRTEM images and size distributions as well as the SAED patterns of the 3DOM $La_{0.6}Sr_{0.4}MnO_3$ and *y*Ag/3DOM $La_{0.6}Sr_{0.4}MnO_3$ samples. The pore sizes of the 3DOM $La_{0.6}Sr_{0.4}MnO_3$ support in *y*Ag/3DOM $La_{0.6}Sr_{0.4}MnO_3$ are 105 ± 10 nm, and the voids are interconnected through open windows with a diameter of 43 × 57 nm, which is in agreement with the results obtained by SEM images (Fig. 5.3). The walls of the macroporous samples are polycrystalline according to the multiple bright electron diffraction rings in the SAED patterns (insets of Fig. 5.4d, h, m, and r). The characteristic SAED rings of $La_{0.6}Sr_{0.4}MnO_3$ are indexed as the (012), (110), (202), (116), (300), (208), and (128) lattice planes of the face-centered-cubic structure. In addition, the spherical Ag NPs deposited on the surface of 3DOM $La_{0.6}Sr_{0.4}MnO_3$ are clearly observed in the HRTEM images (Figs. A.11 and A.12 of Appendix), on which all of the Ag NPs are highly dispersed and uniform in size without any agglomerations, as indicated in the dotted cycles of Fig. 5.4j–m.

With the rise in Ag NPs loading to 5.71 wt%, however, a great tendency to agglomerate was observed (Fig. A.13 of Appendix), and the particle size of Ag NPs on the macropore skeletons increased (Fig. 5.4o–r). The magnification of Ag NPs in Fig. 5.4h, m, and r shows that the spherical silver NPs locate on the marcopore channels of $La_{0.6}Sr_{0.4}MnO_3$ (Fig. 5.3j and k). The size distributions of Ag NPs in

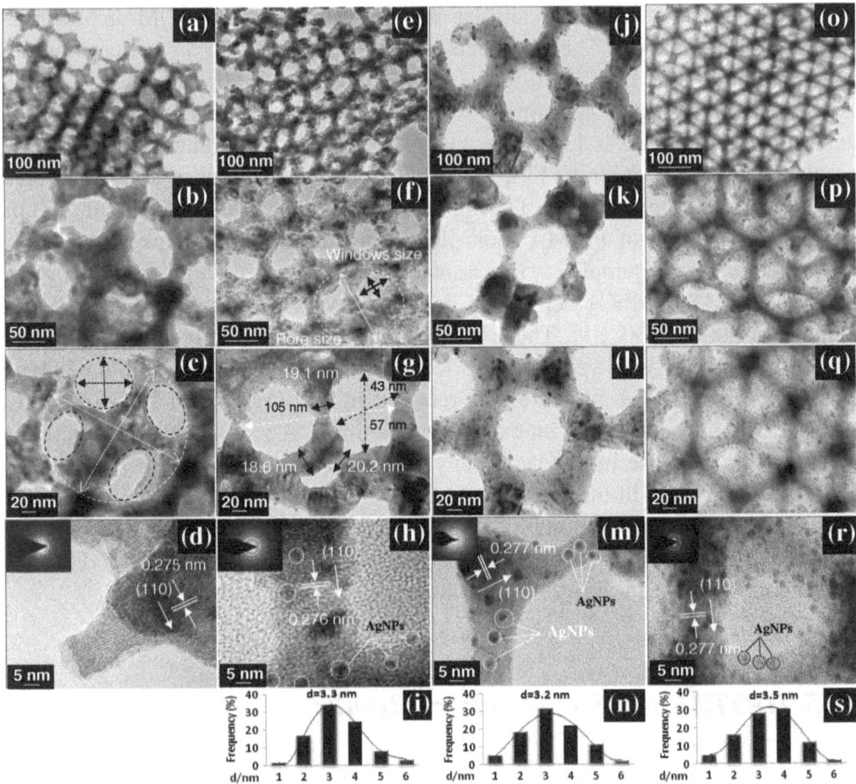

Fig. 5.4 TEM and high-resolution TEM images as well as the SAED patterns (*insets*) of **a–d** 3DOM $La_{0.6}Sr_{0.4}MnO_3$, **e–i** 1.58 wt% Ag/3DOM $La_{0.6}Sr_{0.4}MnO_3$, **j–n** 3.63 wt% Ag/3DOM $La_{0.6}Sr_{0.4}MnO_3$, and **o–s** 5.71 wt% Ag/3DOM $La_{0.6}Sr_{0.4}MnO_3$

*y*Ag/3DOM $La_{0.6}Sr_{0.4}MnO_3$ are in the range of 2.9–6.9 nm (Fig. 5.4i, n, and s and Table 5.1), and the mean diameter of Ag NPs is 3.3 nm. It is quite clear that the Ag NPs in 3.63 wt% Ag/3DOM $La_{0.6}Sr_{0.4}MnO_3$ are smaller in size and more uniform in size distribution as compared to those in 5.71 wt% Ag/3DOM $La_{0.6}Sr_{0.4}MnO_3$. We suggest that the formation of nanovoids or nanopores on the walls of 3DOM $La_{0.6}Sr_{0.4}MnO_3$ could be related to the presence of DMOTEG and PEG400 during the preparation process as well as the strong metal-support interactions (SMSIs). The interaction between the Ag NPs and support would be favorable for the enhancement in mobility of lattice oxygen and may be conducive to the production of oxygen vacancies on the surface of *y*Ag/3DOM $La_{0.6}Sr_{0.4}MnO_3$ [39, 40]. Therefore, the synergistic effect of nano-composite between nanometric crystal of the $La_{0.6}Sr_{0.4}MnO_3$ support and Ag NPs is favorable for enhancing the redox ability of the catalysts.

5.2.6 Pore Structure and Surface Area (BET)

The N_2 adsorption–desorption isotherms and pore-size distributions of the 1D $La_{0.6}Sr_{0.4}MnO_3$, 3DOM $La_{0.6}Sr_{0.4}MnO_3$, and *y*Ag/3DOM $La_{0.6}Sr_{0.4}MnO_3$ samples are shown in Fig. 5.5. Each of the samples displayed a type II nitrogen adsorption–desorption isotherms with a type H3 hysteresis loop in the relative pressure (p/p_0) range of 0.9–1.0. The hysteresis loops in the low and high p/p_0 ranges of the 3DOM $La_{0.6}Sr_{0.4}MnO_3$ and *y*Ag/3DOM $La_{0.6}Sr_{0.4}MnO_3$ samples were slightly different from that of the 1D $La_{0.6}Sr_{0.4}MnO_3$ sample, indicative of the discrepancy in pore-size distribution (Fig. 5.5a). For the low-pressure portion, a near linear middle section of each isotherm was observed, which indicates that there was a co-presence of macropores (in majority) and nanovoids or mesopores (in minority) of the *y*Ag/3DOM $La_{0.6}Sr_{0.4}MnO_3$ samples. The nanovoids or mesopores contribute to the high BET surface area. It can be clearly shown in Fig. 5.5a that the hysteresis loop of *y*Ag/3DOM $La_{0.6}Sr_{0.4}MnO_3$ is located in the p/p_0 range of 0.45–1.0, and it is much bigger than that of 3DOM $La_{0.6}Sr_{0.4}MnO_3$.

The BET surface areas (38.2–42.7 m^2/g) of the *y*Ag/3DOM $La_{0.6}Sr_{0.4}MnO_3$ samples were significantly higher than that (2.6 m^2/g) of the 1D $La_{0.6}Sr_{0.4}MnO_3$ sample. The pore volumes of these porous materials were in the range of 0.159–0.165 cm^3/g. Clearly, the introduction of DMOTEG and loading of Ag NPs during the fabrication process are favorable for the enhancement in surface area of the as-fabricated samples.

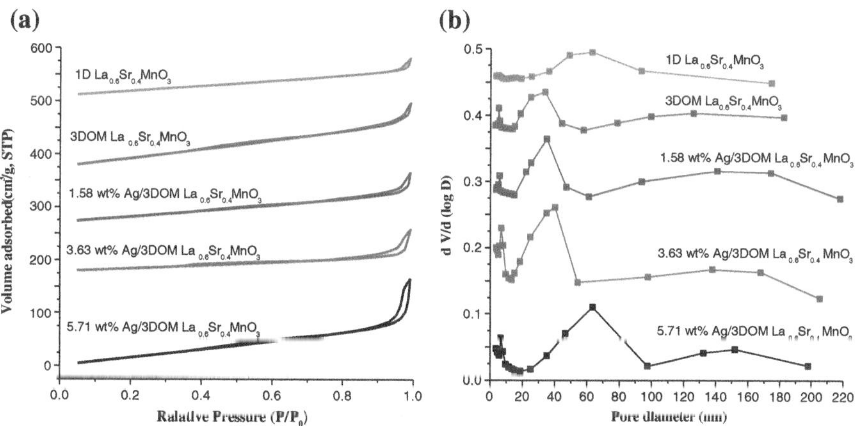

Fig. 5.5 **a** Nitrogen adsorption–desorption isotherms and **b** pore-size distributions of the 1D $La_{0.6}Sr_{0.4}MnO_3$, 3DOM $La_{0.6}Sr_{0.4}MnO_3$, and *y*Ag/3DOM $La_{0.6}Sr_{0.4}MnO_3$ samples

5.2.7 X-ray Photoelectron Spectroscopy (XPS)

Figure 5.6 illustrates the Mn $2p_{3/2}$, Ag $3d_{5/2}$, and O 1s XPS spectra of 1D $La_{0.6}Sr_{0.4}MnO_3$ and *y*Ag/3DOM $La_{0.6}Sr_{0.4}MnO_3$ samples. It can be observed from Fig. 5.6a that the asymmetrical Mn $2p_{3/2}$ XPS spectrum of each sample could be decomposed to two components at BE = 641.9 and 644.1 eV, assignable to the surface of Mn^{3+} and Mn^{4+} species [41], respectively. As summarized in Table 5.2, with the rise in Ag loading, the surface Mn^{4+}/Mn^{3+} molar ratios of the *y*Ag/3DOM $La_{0.6}Sr_{0.4}MnO_3$ samples first increased, with the 3.63 wt% Ag/3DOM $La_{0.6}Sr_{0.4}MnO_3$ sample showing the highest surface Mn^{4+}/Mn^{3+} molar ratio (1.52). It is found that the surface La/Mn (0.94–1.04) and La/(Mn + La) (0.46–0.51) molar ratios of *y*Ag/3DOM $La_{0.6}Sr_{0.4}MnO_{3,}$ are lower than the corresponding nominal La/Mn (1.00) and Mn/(Mn + La) (0.20) molar ratios (Table 5.2), suggesting the presence of Mn enrichment on the surface of the three samples. The Ag $3d_{5/2}$ XPS spectra of the 1D $La_{0.6}Sr_{0.4}MnO_3$ and 1.58–3.63 wt% Ag/3DOM $La_{0.6}Sr_{0.4}MnO_3$ samples are shown in Fig. 5.6b, in which a main peak at BE = 368.3 and 374.2 eV is assigned to the Ag^0 and $Ag^{\delta+}$ species, respectively. The $Ag^{\delta+}/Ag^0$ molar ratios of the *y*Ag/3DOM $La_{0.6}Sr_{0.4}MnO_3$ samples increased with the rise in Ag loading 0.25 for the 5.71 wt% Ag/3DOM $La_{0.6}Sr_{0.4}MnO_3$ sample, whereas 0.19 for the 1.58 wt% Ag/3DOM $La_{0.6}Sr_{0.4}MnO_3$ sample. The ion states of Ag species ($Ag^{\delta+}$) are considered to be more active than the metal state (Ag^0) for oxidation reaction [42].

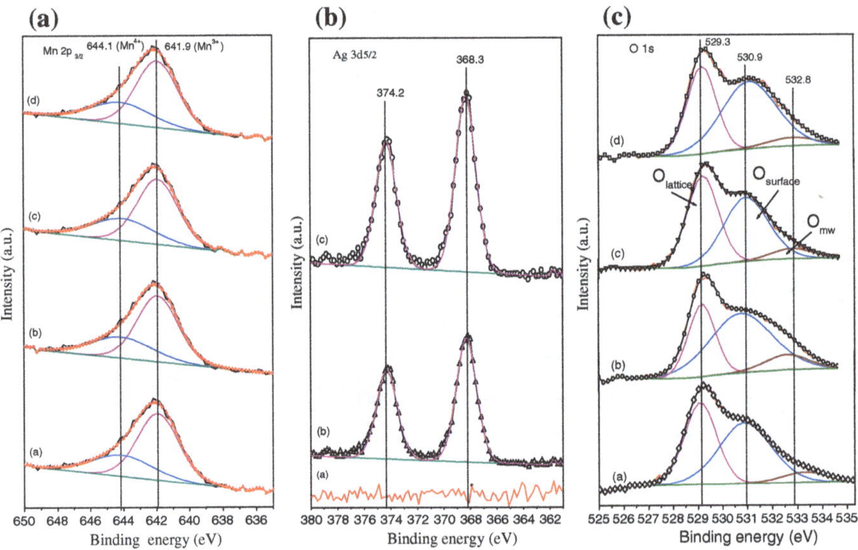

Fig. 5.6 **a** Mn $2p_{3/2}$, **b** Ag $3d_{5/2}$, and **c** O 1s XPS spectra of (*a*) 1D $La_{0.6}Sr_{0.4}MnO_3$, (*b*) 1.58 wt% Ag/3DOM $La_{0.6}Sr_{0.4}MnO_3$, (*c*) 3.63 wt% Ag/3DOM $La_{0.6}Sr_{0.4}MnO_3$, and (*d*) 5.71 wt% Ag/3DOM $La_{0.6}Sr_{0.4}MnO_3$

Table 5.2 Surface element compositions, H_2 consumptions, and catalytic activities of the 1D $La_{0.6}Sr_{0.4}MnO_3$, 3DOM $La_{0.5}Sr_{0.4}MnO_3$, and *y*Ag/3DOM $La_{0.6}Sr_{0.4}MnO_3$ samples

Catalyst	Ag	La				Mn				O					H_2 consumption[e] (mmol/g)		Methane combustion[f] (°C)		
	$Ag^{\delta+}/Ag^0$	3d		La/Mn	La/(Mn + La)	2p/32		Mn^{4+}/Mn^{3+}	Mn/(La + Mn + O)	H_2O	O_{lat}^c		O_{ads}^d	O_{ads}/O_{lat}^{2-}	50–550 °C	550–950 °C	$T_{10\%}$	$T_{50\%}$	$T_{90\%}$
	Molar ratio[a]	BE[a]		Molar ratio	Molar ratio	BE		Molar ratio	Molar ratio	BE	BE	Area	BE	Molar ratio					
1D-$La_{0.6}Sr_{0.4}MnO_3$	–/–	834.4	837.9	1.01 (1.00)	0.51 (0.50)	641.9	644.1	0.91	0.18 (0.20)	532.8	529.3	1200	530.9	0.81	1.43	1.52	562	672	749
3DOM $La_{0.6}Sr_{0.4}MnO_3$	–/–	834.3	837.9	1.04 (1.00)	0.50 (0.50)	641.8	644.1	1.45	0.19 (0.20)	532.9	529.3	1067	530.8	1.22	1.86	1.49	437	566	661
1.58 wt% Ag/3DOM $La_{0.6}Sr_{0.4}MnO_3$	0.19	834.6	838.0	0.94 (1.00)	0.46 (0.50)	641.6	644.2	1.48	0.18 (0.20)	532.8	529.2	1298	530.8	1.83	1.92	1.40	436	565	661
3.63 wt% Ag/3DOM $La_{0.6}Sr_{0.4}MnO_3$	0.22	834.3	837.8	0.97 (1.00)	0.48 (0.50)	641.7	644.1	1.52	0.17 (0.20)	532.9	529.2	1340	530.9	2.01	1.94	1.44	361	454	524
5.71 wt% Ag/3DOM $La_{0.6}Sr_{0.4}MnO_3$	0.25	834.5	837.7	0.99 (1.00)	0.49 (0.50)	641.7	644.2	1.47	0.16 (0.20)	532.8	529.4	1276	530.7	1.69	1.92	1.45	399	498	697

[a]Data in parenthesis are the nominal La/Mn and/or Mn/(La + Mn + O) molar ratios
[b]Binding energy (eV)
[c]Lattice oxygen species
[d]Adsorbed oxygen species
[e]Data based on quantitative analysis of H_2-TPR profiles
[f]Reaction conditions: 2 % CH_4 + 20 % O_2 + 78 % N_2 (balance), total flow = 41.6 ml/min, GHSV = ca. 30,000 ml/(g h)

Hence, these deconvolutions allow us to estimate the relative amount of each Ag species (Ag^0 and $Ag^{\delta+}$) on the surface of the *y*Ag/3DOM $La_{0.6}Sr_{0.4}MnO_3$ samples, and the results are presented in Table 5.2. For each sample, the O 1s spectrum (Fig. 5.6c) could be decomposed to three components at BE = 529.3, 530.9, and 532.8 eV, attributable to the surface lattice oxygen (O_{lat}), adsorbed oxygen (O_{ads}, e.g., O_2^-, O_2^{2-} or O^-), and carbonate species [43], respectively. As summarized in Table 5.2, the surface O_{ads}/O_{lat} molar ratio (1.83–2.01) increased remarkably after the loading of Ag NPs on the $La_{0.6}Sr_{0.4}MnO_3$ support. The 3.63 wt% Ag/3DOM $La_{0.6}Sr_{0.4}MnO_3$ and 1D $La_{0.6}Sr_{0.4}MnO_3$ samples show the highest (2.01) and lowest (0.81) O_{ads}/O_{lat} molar ratios, respectively. It is known that O_2, O_2^{2-} or O^- species are very active for the oxidation of hydrocarbons [31, 44]. The surface O_{ads}/O_{lat} molar ratios (Table 5.2) decreased in the order of 3.63 wt% Ag/3DOM $La_{0.6}Sr_{0.4}MnO_3$ (2.01) > 1.58 wt% Ag/3DOM $La_{0.6}Sr_{0.4}MnO_3$ (1.83) > 5.71 wt%.

Ag/3DOM $La_{0.6}Sr_{0.4}MnO_3$ (1.69) > 3DOM $La_{0.6}Sr_{0.4}MnO_3$ (1.22) > 1D $La_{0.6}Sr_{0.4}MnO_3$ (0.81). The rise in surface active oxygen species concentration would give rise to enhanced catalytic performance of 3.63 wt% Ag/3DOM $La_{0.6}Sr_{0.4}MnO_3$ for the combustion of methane.

5.2.8 Reducibility (H_2-TPR)

H_2-TPR experiments were conducted to investigate the reducibility of the 1D $La_{0.6}Sr_{0.4}MnO_3$, 3DOM $La_{0.6}Sr_{0.4}MnO_3$, and *y*Ag/3DOM $La_{0.6}Sr_{0.4}MnO_3$ samples, and their profiles are illustrated in Fig. 5.7 and Table 5.3. As shown in Fig. 5.7a, two reduction peaks at 456 and 723 °C are observed for the 3DOM $La_{0.6}Sr_{0.4}MnO_3$ sample. Since La^{3+} and Sr^{2+} both are nonreducible under the H_2-TPR conditions adopted in this study, the observed reduction peaks are due to the reduction of Mn^{n+} species [28, 31]. When Ag NPs were loaded onto the $La_{0.6}Sr_{0.4}MnO_3$ surface, all of the reduction peaks shifted to lower temperatures. For the 3.63 wt% Ag/3DOM $La_{0.6}Sr_{0.4}MnO_3$ sample, the high-temperature reduction peak at 721 °C leading to the breakdown of the perovskite phase was almost unchanged in position and shape, as compared to that of the Ag-free 3DOM $La_{0.6}Sr_{0.4}MnO_3$ sample (Fig. 5.7a and b). The above XPS results (Fig. 5.6b) have shown that there still existed an appreciable quantity of Ag^+ species on the surface of *y*Ag/3DOM $La_{0.6}Sr_{0.4}MnO_3$, and that the modification of Ag resulted in formation of a certain quantity of Mn^{4+}. Thus, the small peak at 304 °C might be assigned to the reduction of Ag^+ species [45]. By comparing the reduction profiles of the 3.63 wt% Ag/3DOM $La_{0.6}Sr_{0.4}MnO_3$ and 3DOM $La_{0.6}Sr_{0.4}MnO_3$ samples (Fig. 5.7b), one can see that the porous structure promoted the dispersion of Ag NPs, thus favoring the reduction of Mn species. For the 3.63 wt% Ag/3DOM $La_{0.6}Sr_{0.4}MnO_3$ and 1D $La_{0.6}Sr_{0.4}MnO_3$ samples, the H_2 consumption at low temperatures (<550 °C) was 1.92–1.93 and 1.43 mmol/g, while that at high temperatures it (>550 °C) was

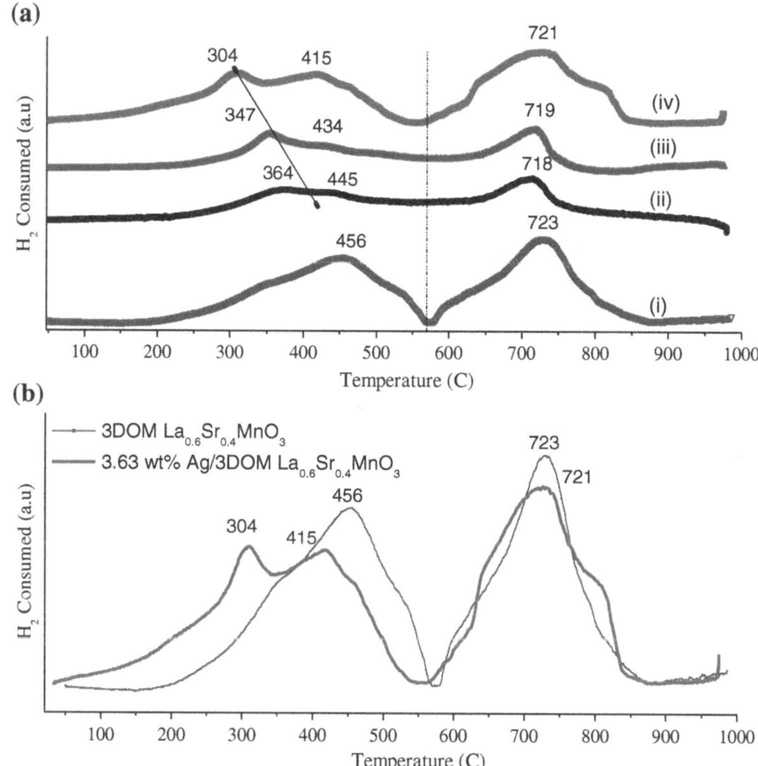

Fig. 5.7 **a**, **b** H_2-TPR profiles (*i* 3DOM $La_{0.6}Sr_{0.4}MnO_3$; *ii* 1.58 wt% Ag/3DOM $La_{0.6}Sr_{0.4}MnO_3$; *iii* 3.63 wt% Ag/3DOM $La_{0.6}Sr_{0.4}MnO_3$; *iv* 5.71 wt% Ag/3DOM $La_{0.6}Sr_{0.4}MnO_3$)

1.40–146 and 1.52 mmol/g, respectively. The total H_2 consumption of 3DOM $La_{0.6}Sr_{0.4}MnO_3$ and *y*Ag/3DOM $La_{0.6}Sr_{0.4}MnO_3$ samples was in the range of 2.95–3.38 mmol/g (Table 5.2 and Fig. 5.8). Figure 5.8 shows the initial H_2 consumption rate as a function of inverse temperature of the 1D $La_{0.6}Sr_{0.4}MnO_3$, 3DOM $La_{0.6}Sr_{0.4}MnO_3$, and *y*Ag/3DOM $La_{0.6}Sr_{0.4}MnO_3$ samples. It is clearly seen that the initial H_2 consumption rate was decreased in the order of 3.63 wt% Ag/3DOM $La_{0.6}Sr_{0.4}MnO_3$ < 5.71 wt% Ag/3DOM $La_{0.6}Sr_{0.4}MnO_3$ < 3DOM $La_{0.6}Sr_{0.4}MnO_3$ < 1.58 wt% Ag/3DOM $La_{0.6}Sr_{0.4}MnO_3$ < 1D $La_{0.6}Sr_{0.4}MnO_3$.

The trend in low-temperature reducibility was in good agreement with the sequence in catalytic activity shown below. Through the above comparison, we believe that Ag NPs greatly improves the low-temperature reducibility of the $La_{0.6}Sr_{0.4}MnO_3$ support, remarkably increasing the content of low-temperature-reducible Mn^{n+} species.

Table 5.3 Turnover frequencies (TOFs), rate constants (k), reaction rates (r), activation energies (E_a), and correlation coefficients (R^2) of the 1D $La_{0.6}Sr_{0.4}MnO_3$, 3DOM $La_{0.6}Sr_{0.4}MnO_3$, and yAg/3DOM $La_{0.6}Sr_{0.4}MnO_3$ samples for methane combustion at different temperatures

Catalyst	TOF_{Ag} (mol/mol_{Ag} s)			TOF_{Mn} (mol/mol_{Mn} s)			k ($\times 10^{-3}$ s^{-1})/r[a] ($\times 10^{-4}$ μmol/(g s))[b]				Kinetic parameter	
	300 °C	350 °C	400 °C	300 °C	350 °C	400 °C	300 °C	400 °C	500 °C	550 °C	E_a (kJ/mol)	R^2
1D $La_{0.6}Sr_{0.4}MnO_3$	–/–	–/–	–/–	1.73×10^{-8}	5.73×10^{-8}	8.40×10^{-8}	0.16/0.37	0.25/0.55	0.35/0.76	0.42/0.92	91.4	0.9991
3DOM $La_{0.6}Sr_{0.4}MnO_3$	–/–	–/–	–/–	4.17×10^{-7}	6.22×10^{-7}	7.92×10^{-7}	0.67/1.43	0.95/1.97	1.31/2.63	2.64/4.71	56.6	0.9990
1.58 wt% Ag/3DOM $La_{0.6}Sr_{0.4}MnO_3$	1.23×10^{-5}	4.32×10^{-5}	7.28×10^{-5}	5.07×10^{-7}	7.14×10^{-7}	11.1×10^{-7}	1.14/2.56	2.19/4.08	10.7/17.7	37.8/58.7	39.1	0.9995
3.63 wt% Ag/3DOM $La_{0.6}Sr_{0.4}MnO_3$	1.86×10^{-5}	5.68×10^{-5}	11.8×10^{-5}	5.56×10^{-7}	9.03×10^{-7}	16.7×10^{-7}	1.57/3.51	5.86/8.65	13.1/21.0	41.0/61.8	37.5	0.9992
5.71 wt% Ag/3DOM $La_{0.6}Sr_{0.4}MnO_3$	1.31×10^{-5}	2.62×10^{-5}	6.14×10^{-5}	5.05×10^{-7}	8.11×10^{-7}	13.2×10^{-7}	1.33/2.42	2.23/4.69	11.7/15.3	38.2/59.3	38.2	0.9998

[a]Data before the "/" symbol are the k values, whereas those after the "/" symbol are the r values
[b]Reaction condition: 2 % CH_4 + 20 % O_2 + 78 % N_2 (balance), total flow = 41.6 ml/min, GHSV = ca. 30,000 ml/(g h)

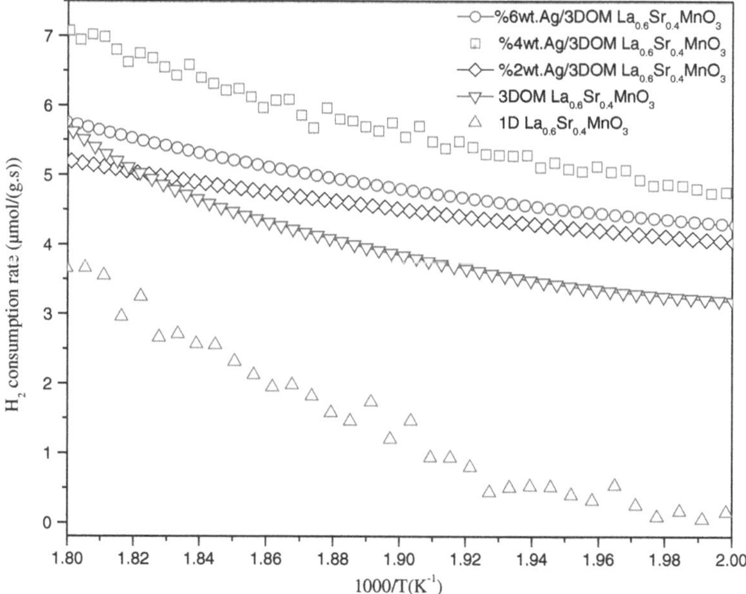

Fig. 5.8 Initial H_2 consumption rate as a function of inverse temperature of the 1D $La_{0.6}Sr_{0.4}MnO_3$, 3DOM $La_{0.6}Sr_{0.4}MnO_3$, and *y*Ag/3DOM $La_{0.6}Sr_{0.4}MnO_3$ samples

5.3 Activity Evaluation of Catalyst

5.3.1 *Influence of Different Ag Loading on the Activity of Catalyst*

We prepared the 3DOM $La_{0.6}Sr_{0.4}MnO_3$ catalysts and investigated the catalytic performance of 3DOM $La_{0.6}Sr_{0.4}MnO_3$-supported Ag NPs for methane combustion. As shown in Fig. 5.9 and Table 5.2, the catalytic activities of 1D $La_{0.6}Sr_{0.4}MnO_3$ and *y*Ag/3DOM $La_{0.6}Sr_{0.4}MnO_3$ samples increased with the rise in temperature. The highest activity was achieved due to both the smaller size of Ag NPs and larger surface area of 3.63 wt% Ag/3DOM $La_{0.6}Sr_{0.4}MnO_3$. As expected, the catalytic activity decreased at elevated GHSV values. At GHSV = 30,000 ml/(g h), the $T_{50\%}$ and $T_{90\%}$ were 454 and 524 °C, respectively, which were 87 and 113 °C lower than those achieved at GHSV = 40,000 ml/(g h). A further rise in GHSV from 40,000 to 45,000 ml/(g h) resulted in a drop in catalytic activity. The Ag particle size is in the range of 1–6 nm with a narrow distribution and the mean diameters of the *y*Ag/3DOM $La_{0.6}Sr_{0.4}MnO_3$ (y = 1.58, 3.63, and 5.71 wt%) samples are 3.4, 3.3, and 3.5 nm (Fig. 5.4i, n, and s), respectively. It is convenient to compare the catalytic activities of the samples using the reaction temperatures $T_{10\%}$, $T_{50\%}$, and $T_{90\%}$ (corresponding to methane conversion = 10, 50, and 90 %), as summarized in Table 5.2.

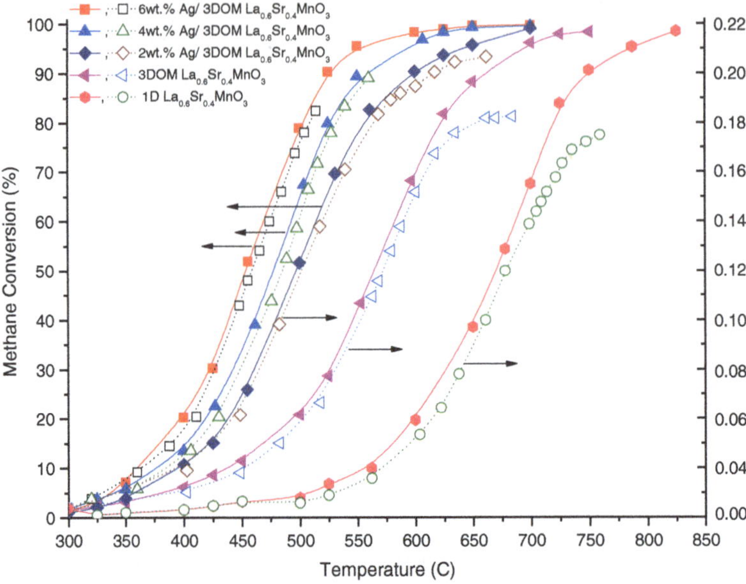

Fig. 5.9 Methane conversion and the corresponding reaction rate versus reaction temperature over the 1D $La_{0.6}Sr_{0.4}MnO_3$, 3DOM $La_{0.6}Sr_{0.4}MnO_3$, and *y*Ag/3DOM $La_{0.6}Sr_{0.4}MnO_3$ catalysts

It can be found that the catalytic performance decreased in the sequence of 3.63 wt% Ag/3DOM $La_{0.6}Sr_{0.4}MnO_3$ > 5.71 wt% Ag/3DOM $La_{0.6}Sr_{0.4}MnO_3$ > 1.58 wt% Ag/3DOM $La_{0.6}Sr_{0.4}MnO_3$ > 3DOM $La_{0.6}Sr_{0.4}MnO_3$ > 1D $La_{0.6}Sr_{0.4}MnO_3$, coinciding with the order in low-temperature reducibility (i.e., the initial H_2 consumption rate (Fig. 5.8) of the catalysts). It can be found that during 24 h on stream reaction, no large drop in catalytic performance was studied (Fig. A.14 of Appendix). The 3D porous structure was catalytically steady. This summary was proved by the effects of HRSEM and XRD studies (Fig. A.15 of Appendix), in which no obvious change in 3DOM material and crystal phase was detected.

Obviously, the 3.63 wt% Ag/3DOM $La_{0.6}Sr_{0.4}MnO_3$ catalyst with a surface area of 41.5 m^2/g performed the best, giving the $T_{10\%}$, $T_{50\%}$, and $T_{90\%}$ of 361, 454, and 524 °C, respectively, which were much lower than those ($T_{10\%}$ = 562 °C, $T_{50\%}$ = 672 °C, and $T_{90\%}$ = 749 °C) achieved over the 1D $La_{0.6}Sr_{0.4}MnO_3$ sample with a surface area of 2.6 m^2/g by 201, 216, and 225 °C, respectively. According to the activity data and mole of Mn in the $La_{0.6}Sr_{0.4}MnO_3$ and *y*Ag/3DOM $La_{0.6}Sr_{0.4}MnO_3$ catalysts, we calculated the turnover frequency (TOF_{Ag} and TOF_{Mn}) and reaction rates (μmol/(g s)) according to the single surface Ag site (or mole of Mn) and the Ag weight in the *y*Ag/$La_{0.6}Sr_{0.4}MnO_3$ samples, respectively. The TOF_{Ag} (mol/mol_{Ag} s) was calculated according to the number of CH_4 molecules converted by single surface Ag site per second. The TOF_{Ag} and TOF_{Mn} of the samples were calculated under the conditions of CH_4/O_2 molar ratio = 1/10,

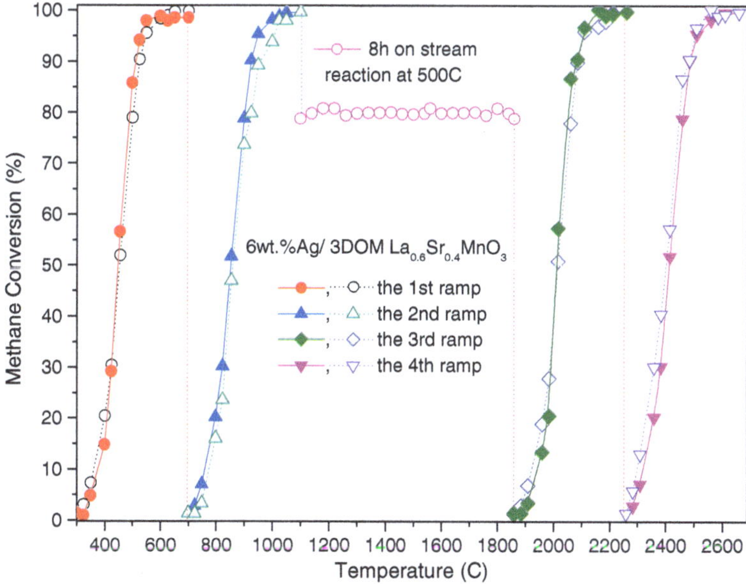

Fig. 5.10 Catalytic stability of the 3.63 wt% Ag/3DOM $La_{0.6}Sr_{0.4}MnO_3$ catalyst at GHSV = 30,000 ml/(g h) under the conditions of 2 % CH_4 + 20 % O_2 + 78 % N_2 (balance) and the total flow of 41.6 ml/min

GHSV = 30,000 ml/(g h), and temperature = 300, 350, and 400 °C, as summarized in Table 5.3 (Fig. 5.10).

It is observed that in the nonporous 1D $La_{0.6}Sr_{0.4}MnO_3$ sample at the same temperature, the TOF_{Mn} values of the porous 3DOM $La_{0.6}Sr_{0.4}MnO_3$ and *y*Ag/3DOM $La_{0.6}Sr_{0.4}MnO_3$ samples were much higher. For instance, the TOF_{Mn} value (5.56×10^{-7}) of 3.63 wt% Ag/3DOM $La_{0.6}Sr_{0.4}MnO_3$ was approximately 1.5 times as much as that (4.17×10^{-7}) of 3DOM $La_{0.6}Sr_{0.4}MnO_3$ for methane combustion at 300 °C. With the rise in Ag NPs loading from 1.58 to 3.63 wt%, the obtained 1.58 wt% Ag/3DOM $La_{0.6}Sr_{0.4}MnO_3$ and 3.63 wt% Ag/3DOM $La_{0.6}Sr_{0.4}MnO_3$ samples exhibited the TOF_{Mn} values of 5.07×10^{-7} and 5.56×10^{-7} mol/(mol_{Mn} s) at 300 °C, and 7.14×10^{-7} and 9.03×10^{-7} mol/(mol_{Mn} s) at 350 °C, respectively; but they decreased to 5.05×10^{-7} mol/(mol_{Mn} s) at 300 °C and 8.11×10^{-7} mol/(mol_{Mn} s) at 400 °C with a further rise in Ag NPs loading from 3.63 to 5.71 wt%. In the meantime, from Table 5.3, one can also observe that the TOF_{Ag} values (1.86×10^{-5} mol/(mol_{Ag} s) at 300 °C and 11.8×10^{-5} mol/(mol_{Ag} s) at 400 °C) of the 3.63 wt% Ag/3DOM $La_{0.6}Sr_{0.4}MnO_3$ sample were much higher than those (1.23×10^{-5} mol/(mol_{Ag} s) at 300 °C and 7.28×10^{-5} mol/(mol_{Ag} s) at 400 °C) of the 1.58 wt% Ag/3DOM $La_{0.6}Sr_{0.4}MnO_3$ sample. This result indicates that there was a SMSI between the metal (Ag NPs) and the support ($La_{0.6}Sr_{0.4}MnO_3$), which gave rise to an enhanced catalytic performance of *y*Ag/3DOM $La_{0.6}Sr_{0.4}MnO_3$ for methane combustion. In addition, the catalytic stability of the 3.63 wt% Ag/3DOM $La_{0.6}Sr_{0.4}MnO_3$ sample was tested with a test time of 30 h. The result reveals that no obvious activity loss was detected within 30 h of the reaction. Therefore, it is concluded that the good catalytic performance of 3.63 wt% Ag/3DOM $La_{0.6}Sr_{0.4}MnO_3$ for methane combustion was associated with its higher surface area, higher surface oxygen species concentration, and better low-temperature reducibility as well as the good-quality nanovoid-walled 3DOM structure.

5.3.2 Influence of H_2O and SO_2 on the Activity of Catalyst

We conducted the oxidation of methane in the presence of 1.0, 3.0, or 5.0 vol% of water vapor or 40 ppm SO_2 over the 3.63 wt% Ag/3DOM $La_{0.6}Sr_{0.4}MnO_3$ and $La_{0.6}Sr_{0.4}MnO_3$ samples at different temperatures, and the results are shown in Figs. 5.11 and 5.12. Feed gas-containing water vapor was introduced into a catalyst bed at a temperature of 550, 600, or 650 °C for 1 h, and then water-free gas was fed for 1 h. This alternating cycle was repeated for four times. When the catalytic activity reached a steady value, 3.0 vol% of water vapor in the feed was introduced to the reaction system. It is observed from Fig. 5.11a that the addition of water vapor at 550 or 600 °C decreased methane conversion by ca. 15–20 %. Methane conversion after onstream reaction at 550 °C for 10 h was 66.4 %, which was lower than that (77.9 %) after onstream reaction at 550 °C for 10 min. Therefore, the addition of 3.0 vol% of water vapor did not affect the catalytic activity of the 3.63 wt% Ag/3DOM $La_{0.6}Sr_{0.4}MnO_3$ sample if the reaction temperature was higher than 600 °C, but it decreased the activity when the reaction temperature was below 550 °C.

Meanwhile, the influence of water vapor concentration (1.0, 3.0 or 5.0 vol%) on the activity of the 3.63 wt% Ag/3DOM $La_{0.6}Sr_{0.4}MnO_3$ catalyst for methane

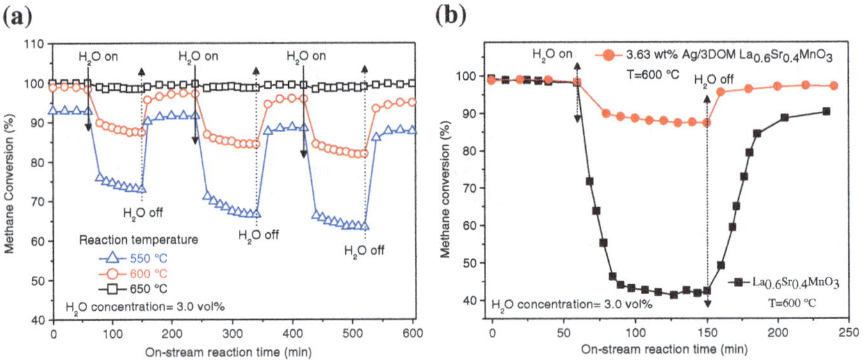

Fig. 5.11 Effect of water vapor on methane conversion at different reaction temperatures over the 3.63 wt% Ag/3DOM $La_{0.6}Sr_{0.4}MnO_3$ catalyst (H_2O concentration = 3.0 vol%), **b** water introduction or cutting off in the feedstock over 3.63 wt% Ag/3DOM $La_{0.6}Sr_{0.4}MnO_3$ and $La_{0.6}Sr_{0.4}MnO_3$ at 600 °C (H_2O concentration = 3.0 vol%)

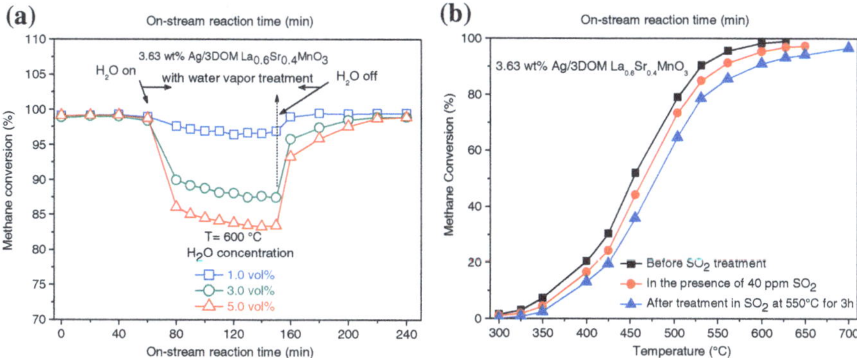

Fig. 5.12 Effect of water vapor concentration, and **b** effect of SO_2 on methane conversion over the 3.63 wt% Ag/3DOM $La_{0.6}Sr_{0.4}MnO_3$ catalyst under the conditions of GHSV = 30,000 ml/(g h) and SO_2 concentration = 40 ppm

oxidation is shown in Fig. 5.12a. The catalytic activity was not significantly affected at a lower water vapor concentration (1.0 vol%), whereas introduction of a higher water vapor concentration (5.0 vol%) decreased the $T_{90\%}$ value by ca. 12 %.

Furthermore, the HRTEM images, BET surface area, and also XRD pattern of the used catalyst in the case of addition of 3.0 vol% of water vapor was rather similar to that of the fresh one, and the surface area (40.6 m^2/g) of the former was close to that (41.5 m^2/g) of the latter. In order to examine the effect of SO_2 on catalytic activity of the 3.63 wt% Ag/3DOM $La_{0.6}Sr_{0.4}MnO_3$ sample, we carried out methane oxidation in the presence of 40 ppm SO_2 in the feedstock and then switched the SO_2-containing atmosphere to SO_2-free atmosphere at 550 °C (Fig. 5.12b). However, no significant deactivation due to SO_2 introduction was observed over the 3.63 wt% Ag/3DOM $La_{0.6}Sr_{0.4}MnO_3$ sample after this catalyst was treated in 40 ppm SO_2 at 550 °C for 3 h, and the catalytic activity decreased by ca. 5 %. Furthermore, the XRD pattern (Fig. A.16 of Appendix) of the used catalyst in the case of addition of 3.0 vol% of water vapor was rather similar to that of the fresh one, and the surface area (40.6 m^2/g) of the former was close to that (41.5 m^2/g) of the latter. The HRSEM and HRTEM images (Fig. A.17 of Appendix) of the used sample further reveal that the Ag NPs were well stabilized on the surface of 3DOM $La_{0.6}Sr_{0.4}MnO_3$. This result indicates that the 3.63 wt% Ag/3DOM $La_{0.6}Sr_{0.4}MnO_3$ sample was good in SO_2 resistance. Such a small loss in catalytic activity was a result due to the strong adsorption of SO_2 on the active sites of $La_{0.6}Sr_{0.4}MnO_3$. According to the results of the present investigation and those reported previously, silver could also improve the resistance to sulfur poisoning, mainly by increasing the concentration of the acidic Mn^{4+} ions and weakening the SO_2 adsorption on the catalyst. However, the SO_2 introduced could adsorb on the surface of $La_{0.6}Sr_{0.4}MnO_3$ to generate a small amount of sulfate that led to a small decrease in catalytic activity. This performance makes *y*Ag/3DOM $La_{0.6}Sr_{0.4}MnO_3$ an attractive material and worthy of further investigation.

5.3.3 Study on Activation Energy of the Catalyst

The kinetics of catalytic oxidation of VOCs has been gained much attention. It is reasonable to suppose that the combustion of methane in the presence of excess oxygen (CH_4/O_2 molar ratio = 1/10) would obey a first-order reaction mechanism with respect to methane concentration (c): $r = -kc = (-A\exp(-E_a/RT))c$, where r, k, A, and E_a are the reaction rate μmol/(g s)), rate constant (s^{-1}), pre-exponential factor, and apparent activation energy (kJ/mol), respectively. Figure 5.13 shows the Arrhenius plots for methane combustion at CH_4 conversion <20 % (at which the temperature range was 300–550 °C) over the 1D $La_{0.6}Sr_{0.4}MnO_3$, 3DOM $La_{0.6}Sr_{0.4}MnO_3$, and yAg/3DOM $La_{0.6}Sr_{0.4}MnO_3$ catalysts. According to the slopes of the well-linear (correlation coefficients (R^2) were above 0.99) Arrhenius plots, one can calculate the k and E_a values for CH_4 combustion over these catalysts, as summarized in Table 5.3. It is clearly seen that the E_a values (91.4 and 56.6 kJ/mol, respectively) of the 1D $La_{0.6}Sr_{0.4}MnO_3$ and 3DOM $La_{0.6}Sr_{0.4}MnO_3$ catalysts were much higher than that (39.1 kJ/mol) of the 1.58 wt% Ag/3DOM $La_{0.6}Sr_{0.4}MnO_3$ catalyst; with the rise in Ag loading from 1.58 to 3.63 wt%, the E_a significant mass transfer limitations were existed in our catalytic system. It is clearly seen that the E_a values (91.4 and 56.6 kJ/mol, respectively) of the 1D $La_{0.6}Sr_{0.4}MnO_3$ and 3DOM $La_{0.6}Sr_{0.4}MnO_3$ catalysts were much higher than that (39.1 kJ/mol) of the 1.58 wt% Ag/3DOM $La_{0.6}Sr_{0.4}MnO_3$ catalyst; with the rise in Ag loading from 1.58 to 3.63 wt%, the E_a value decreased from 39.1 to 37.5 kJ/mol, but a further rise in Ag loading to 5.71 wt% led to a slight increase in E_a value to 38.2 kJ/mol. The E_a value for methane combustion was decreased in a sequence of 1D $La_{0.6}Sr_{0.4}MnO_3$ > 3DOM $La_{0.6}Sr_{0.4}MnO_3$ > 1.58 wt% Ag/3DOM $La_{0.6}Sr_{0.4}MnO_3$ > 5.71 wt% Ag/3DOM $La_{0.6}Sr_{0.4}MnO_3$ > 3.63 wt% Ag/3DOM $La_{0.6}Sr_{0.4}MnO_3$, with the lower

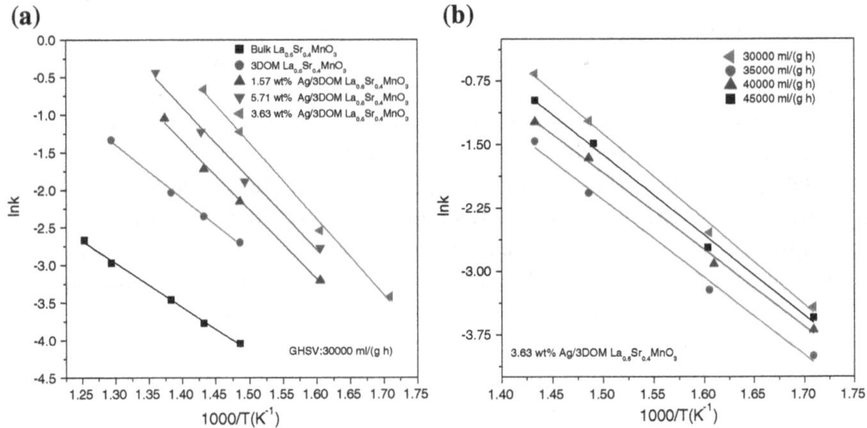

Fig. 5.13 **a** The Arrhenius plots for the methane combustion over the 1D $La_{0.6}Sr_{0.4}MnO_3$, 3DOM $La_{0.6}Sr_{0.4}MnO_3$, and yAg/3DOM $La_{0.6}Sr_{0.4}MnO_3$ samples at GHSV = 30,000 ml/(g h), and **b** 3.63 wt% Ag/3DOM $La_{0.6}Sr_{0.4}MnO_3$ at different GHSV values

E_a values (37.5–39.1 kJ/mol) being achieved over the *y*Ag/3DOM $La_{0.6}Sr_{0.4}MnO_3$ samples (Table 5.3). Such a result suggests that methane oxidation might precede more readily over the porous *y*Ag/3DOM $La_{0.6}Sr_{0.4}MnO_3$ samples. The striking difference in E_a can be likely related to the difference in the total number of active sites, which is directly related to the extent of exposed Ag and $La_{0.6}Sr_{0.4}MnO_3$ surfaces and the presence of a strong Ag–$La_{0.6}Sr_{0.4}MnO_3$ interaction. Therefore, the results of kinetic investigations confirm that the *y*Ag/3DOM $La_{0.6}Sr_{0.4}MnO_3$ catalysts showed excellent performance for the combustion of methane, and it is a promising catalyst for the combustion of methane in practical applications.

5.4 Conclusion and Discussion

A number of rhombohedrally crystallized 3DOM-structured $La_{0.6}Sr_{0.4}MnO_3$-supported Ag NP catalysts with high surface areas (38.2–42.7 m^2/g) were firstly successfully synthesized by a facile novel reduction method in a DMOTEG solution using PMMA colloidal crystal as template. The *y*Ag/3DOM $La_{0.6}Sr_{0.4}MnO_3$ ($y = 0$, 1.57, 3.63, and 5.71 wt%) possess unique nanovoid-walled 3DOM architecture, and the Ag NPs are well dispersed on the inner walls of the uniform macrospores. It was found that the DMOTEG-mediated route not only produced size-controlled Ag NPs, but also stabilized them against conglomeration without the need for additional stabilizers. 3.63 wt% Ag/3DOM $La_{0.6}Sr_{0.4}MnO_3$ catalysts exhibited the large total pore volume (0.165 cm^3/g) and smaller average size distribution (3.2 nm). Ag deposited on the surface of 3DOM $La_{0.6}Sr_{0.4}MnO_3$ did not affect the pore structure and crystalline phase of catalysts. The metal (Ag)-support ($La_{0.6}Sr_{0.4}MnO_3$) synergetic effect is favorable for the transference of the lattice oxygen in support; oxygen species (e.g., O^{2-}, O^- or O_2^-) have been identified for increasing the amounts of oxygen vacancies and active oxygen species as well as Ag^0 and $Ag^{\delta+}$ species. Among the *y*Ag/3DOM $La_{0.6}Sr_{0.4}MnO_3$ samples, 3.63 wt% Ag/3DOM $La_{0.6}Sr_{0.4}MnO_3$ possessed the highest O_{ads} concentration and the best low-temperature reducibility. The 3.63 wt% Ag/3DOM $La_{0.6}Sr_{0.4}MnO_3$ catalyst performed the best, giving the $T_{10\%}$, $T_{50\%}$, and $T_{90\%}$ of 361, 454, and 524 °C at GHSV = ca. 30,000 ml/(g h), respectively, which were much lower than those achieved over the 1D $La_{0.6}Sr_{0.4}MnO_3$ sample by 201, 216, and 225 °C, respectively. The addition of 3.0 vol% water vapor did not affect critically the catalytic activity of the 3.63 wt% Ag/3DOM $La_{0.6}Sr_{0.4}MnO_3$ sample at temperatures above 600 °C, but it decreased the catalytic activity at temperatures below 550 °C. There was a small negative effect of sulfur dioxide on the catalytic activity of the 3.63 wt% Ag/3DOM $La_{0.6}Sr_{0.4}MnO_3$ sample. The 3.63 wt% Ag/3DOM $La_{0.6}Sr_{0.4}MnO_3$ sample exhibited the highest TOF_{Ag} values of 1.86×10^{-5} mol/(mol_{Ag} s) at 300 °C and 11.8×10^{-5} mol/(mol_{Ag} s) at 400 °C. The apparent activation energy (E_a) values (91.4 and 56.6 kJ/mol, respectively) of 1D $La_{0.6}Sr_{0.4}MnO_3$ and 3DOM $La_{0.6}Sr_{0.4}MnO_3$ catalysts were much higher than that (37.5 kJ/mol) of the 3.63 wt% Ag/3DOM $La_{0.6}Sr_{0.4}MnO_3$ catalyst. The 3.63 wt% Ag/3DOM $La_{0.6}Sr_{0.4}MnO_3$

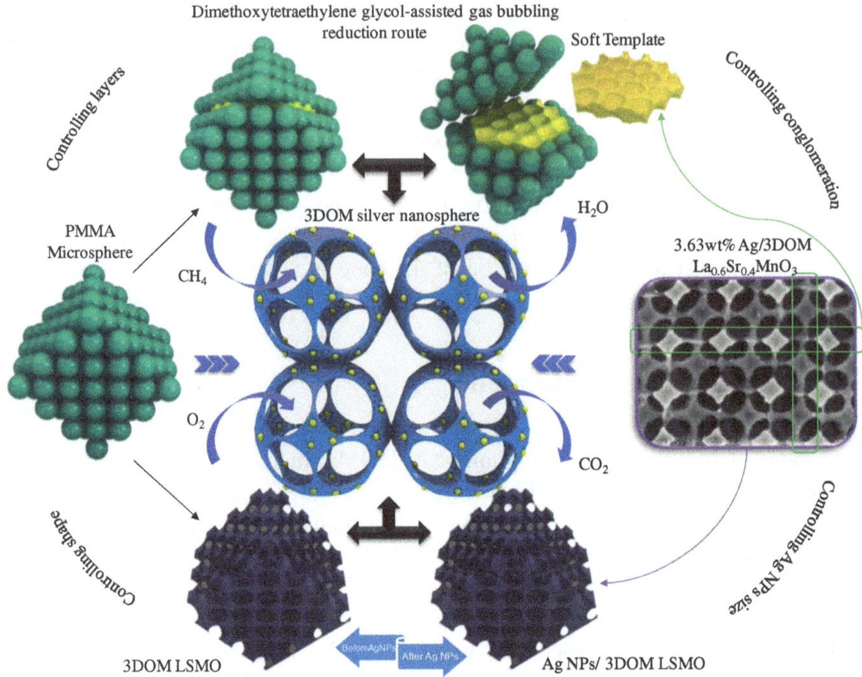

Scheme 5.1 Highly dispersed Ag nanoparticles supported on high-surface-area 3DOM $La_{0.6}Sr_{0.4}MnO_3$ were successfully generated via the dimethoxytetraethylene glycol-assisted gas bubbling reduction route. The macroporous materials showed super catalytic performance for methane combustion

catalyst shows super catalytic activity for methane combustion, which is attributed to a higher oxygen adspecies amount, larger surface area, better low-temperature reducibility, and unique nanovoid-walled 3DOM structure (Scheme 5.1).

References

1. Yi T, Gao S, Qi X, et al. Low temperature synthesis and magnetism of $La_{0.75}Ca_{0.25}MnO_3$ nanoparticles. J Phys Chem Solids. 2000;61(9):1407–13.
2. Tan R, Zhu Y, Feng J, Ji S, Cao L. Preparation of nanosized $LaCo_xMn_{1-x}O_3$ perovskite oxide using amorphous heteronuclear complex as a precursor. J Alloy Compd. 2002;337(1):282–8.
3. Zhu Y, Wang H, Tan R, Cao L. Preparation of nanosized $La_{1-x}Sr_xCoO_3$ via $La_{1-x}Sr_xCo$ (DTPA)·$6H_2O$ amorphous complex precursor. J Alloy Compd. 2003;352(1):134–9.
4. He Y, Zhu Y, Wu N. Synthesis of nanosized $NaTaO_3$ in low temperature and its photocatalytic performance. J Solid State Chem. 2004;177(11):3868–72.
5. Tian Y, He Y, Zhu Y. Low temperature synthesis and characterization of molybdenum disulfide nanotubes and nanorods. Mater Chem Phys. 2004;87(1):87–90.

6. Teng F, Han W, Liang S, Gaugeu B, Zong R, Zhu Y. Catalytic behavior of hydrothermally synthesized $La_{0.5}Sr_{0.5}MnO_3$ single-crystal cubes in the oxidation of CO and CH_4. J Catal. 2007;250(1):1–11.
7. Liang S, Xu T, Teng F, Zong R, Zhu Y. The high activity and stability of $La_{0.5}Ba_{0.5}MnO_3$ nanocubes in the oxidation of CO and CH_4. Appl Catal B. 2010;96(3):267–75.
8. Dokko K, Akutagawa N, Isshiki Y, Hoshina K, Kanamura K. Preparation of three dimensionally ordered macroporous $Li_{0.35}La_{0.55}TiO_3$ by colloidal crystal templating process. Solid State Ionics. 2005;176(31):2345–8.
9. Ding S, Qian W, Tan Y, Wang Y. In-situ incorporation of gold nanoparticles of desired sizes into three-dimensional macroporous matrixes. Langmuir. 2006;22(17):7105–8.
10. Huang J, C-a Tao, An Q, et al. 3D-ordered macroporous poly(ionic liquid) films as multifunctional materials. Chem Commun. 2010;46(6):967–9.
11. Kamio E, Yonemura S, Ono T, Yoshizawa H. Microcapsules with macroholes prepared by the competitive adsorption of surfactants on emulsion droplet surfaces. Langmuir. 2008;24 (23):13287–98.
12. Kotobuki M, Okada N, Kanamura K. Design of a micro-pattern structure for a three dimensionally macroporous Sn-Ni alloy anode with high areal capacity. Chem Commun. 2011;47(21):6144–6.
13. Dai H, Bell AT, Iglesia E. Effects of molybdena on the catalytic properties of vanadia domains supported on alumina for oxidative dehydrogenation of propane. J Catal. 2004;221(2):491–9.
14. Dai H, He H, Li P, Gao L, Au C-T. The relationship of structural defect–redox property–catalytic performance of perovskites and their related compounds for CO and NOx removal. Catal Today. 2004;90(3):231–44.
15. Dai HX, Au CT, Chan Y, Hui KC, Leung YL. Halide-doped perovskite-type $AMn_{1-x}Cu_xO_{3-\delta}$ (A = $La_{0.8}Ba_{0.2}$) catalysts for ethane-selective oxidation to ethene. Appl Catal A. 2001;213 (1):91–102.
16. Dai HX, He H, Au CT. Ethane oxidative dehydrogenation over halogenated $Bi_2Sr_2CaCu_2O_8$-delta catalysts. Ind Eng Chem Res. 2002;41(1):37–45.
17. Choudhary VR, Uphade BS, Pataskar SG, Thite GA. Low-temperature total oxidation of methane over Ag-doped $LaMO_3$ perovskite oxides. Chem Commun. 1996;9:1021–2.
18. Choudhary VR, Uphade BS, Pataskar SG. Low temperature complete combustion of methane over Ag-doped $LaFeO_3$ and $LaFe_{0.5}Co_{0.5}O_3$ perovskite oxide catalysts. Fuel. 1999;78 (8):919–21.
19. Arandiyan H, Chang H, Liu C, Peng Y, Li J. Dextrose-aided hydrothermal preparation with large surface area on 1D single-crystalline perovskite $La_{0.5}Sr_{0.5}CoO_3$ nanowires without template: highly catalytic activity for methane combustion. J Mol Catal A: Chem. 2013;378 (1):299–306.
20. Choudhary VR, Mondal KC. CO_2 reforming of methane combined with steam reforming or partial oxidation of methane to syngas over $NdCoO_3$ perovskite-type mixed metal-oxide catalyst. Appl Energy. 2006;83(9):1024–32.
21. Gardner SD, Hoflund GB, Schryer DR, Schryer J, Upchurch BT, Kielin EJ. Catalytic behavior of noble metal/reducible oxide materials for low-temperature CO oxidation. 1. Comparison of catalyst performance. Langmuir. 1991;7(10):2135–9.
22. Song KS, Kang SK, Kim SD. Preparation and characterization of Ag/MnOx/perovskite catalysts for CO oxidation. Catal Lett. 1997;49(1–2):65–8.
23. Li X, Dai H, Deng J, et al. In situ PMMA-templating preparation and excellent catalytic performance of Co_3O_4/3DOM $La_{0.6}Sr_{0.4}CoO_3$ for toluene combustion. Appl Catal A. 2013;458(1):11–20.
24. Wang Y, Dai H, Deng J, et al. 3DOM InVO4-supported chromia with good performance for the visible-light-driven photodegradation of rhodamine B. Solid State Sci. 2013;24(1):62–70.
25. Wang Y, Dai H, Deng J, et al. Three-dimensionally ordered macroporous InVO4: fabrication and excellent visible-light-driven photocatalytic performance for methylene blue degradation. Chem Eng J. 2013;226(1):87–94.

26. Liu Y, Dai H, Deng J, et al. PMMA-templating generation and high catalytic performance of chain-like ordered macroporous $LaMnO_3$ supported gold nanocatalysts for the oxidation of carbon monoxide and toluene. Appl Catal B. 2013;140(1):317–26.
27. Li X, Dai H, Deng J, et al. Au/3DOM $LaCoO_3$: high-performance catalysts for the oxidation of carbon monoxide and toluene. Chem Eng J. 2013;228(1):965–75.
28. Liu Y, Dai H, Deng J, et al. Au/3DOM $La_{0.6}Sr_{0.4}MnO_3$: highly active nanocatalysts for the oxidation of carbon monoxide and toluene. J Catal. 2013;305(1):146–53.
29. Liu Y, Dai H, Du Y, et al. Controlled preparation and high catalytic performance of three-dimensionally ordered macroporous $LaMnO_3$ with nanovoid skeletons for the combustion of toluene. J Catal. 2012;287(1):149–60.
30. Yuan J, Dai H, Zhang L, et al. PMMA-templating preparation and catalytic properties of high-surface-area three-dimensional macroporous La_2CuO_4 for methane combustion. Catal Today. 2011;175(1):209–15.
31. Arandiyan H, Dai H, Deng J, et al. Three-dimensionally ordered macroporous $La_{0.6}Sr_{0.4}MnO_3$ with high surface areas: active catalysts for the combustion of methane. J Catal. 2013;307(1):327–39.
32. Kucharczyk B, Tylus W. Partial substitution of lanthanum with silver in the $LaMnO_3$ perovskite: effect of the modification on the activity of monolithic catalysts in the reactions of methane and carbon oxide oxidation. Appl Catal A. 2008;335(1):28–36.
33. Ye SL, Song WH, Dai JM, et al. Effect of Ag substitution on the transport property and magnetoresistance of $LaMnO_3$. J Magn Magn Mater. 2002;248(1):26–33.
34. Hien NT, Thuy NP. Preparation and magneto-caloric effect of $La_{1-x}Ag_xMnO_3$ ($x = 0.10$–0.30) perovskite compounds. Phys B. 2002;319(1):168–73.
35. Battabyal M, Dey TK. Low temperature electrical transport in Ag substituted $LaMnO_3$ polycrystalline pellets prepared by a pyrophoric method. Solid State Commun. 2004;131(5):337–42.
36. Ueda W, Sadakane M, Ogihara H. Nano-structuring of complex metal oxides for catalytic oxidation. Catal Today. 2008;132(1):2–8.
37. Caizer C, Stefanescu M. Magnetic characterization of nanocrystalline Ni-Zn ferrite powder prepared by the glyoxylate precursor method. J Phys D Appl Phys. 2002;35(23):3035–40.
38. Sadakane M, Horiuchi T, Kato N, Takahashi C, Ueda W. Facile preparation of three-dimensionally ordered macroporous alumina, iron oxide, chromium oxide, manganese oxide, and their mixed-metal oxides with high porosity. Chem Mater. 2007;19(23):5779–85.
39. Park JB, Graciani J, Evans J, et al. High catalytic activity of Au/CeOx/TiO_2(110) controlled by the nature of the mixed-metal oxide at the nanometer level. Proc Natl Acad Sci USA. 2009;106(13):4975–80.
40. Vayssilov GN, Lykhach Y, Migani A, et al. Support nanostructure boosts oxygen transfer to catalytically active platinum nanoparticles. Nat Mater. 2011;10(4):310–5.
41. Ponce S, Peña MA, Fierro JLG. Surface properties and catalytic performance in methane combustion of Sr-substituted lanthanum manganites. Appl Catal B. 2000;24(3):193–205.
42. Tang W, Hu Z, Wang M, Stucky GD, Metiu H, McFarland EW. Methane complete and partial oxidation catalyzed by Pt-doped CeO_2. J Catal. 2010;273(2):125–37.
43. Machocki A, Ioannides T, Stasinska B, et al. Manganese–lanthanum oxides modified with silver for the catalytic combustion of methane. J Catal. 2004;227(2):282–96.
44. Svensson EE, Nassos S, Boutonnet M, Järås SG. Microemulsion synthesis of MgO-supported $LaMnO_3$ for catalytic combustion of methane. Catal Today. 2006;117(4):484–90.
45. Wang W, Zhang H-B, Lin G-D, Xiong Z-T. Study of Ag/$La_{0.6}Sr_{0.4}MnO_3$ catalysts for complete oxidation of methanol and ethanol at low concentrations. Appl Catal B: Environ. 2000;24(3):219–32.

Chapter 6
Summary

6.1 Conclusions

1. Single-crystalline cubic perovskite mixed oxide $La_{0.5}Sr_{0.5}CoO_3$ nanowire was synthesized via a facile one-pot hydrothermal method with (DHR) route and/or L-Lysine procedure. The BET of nanowire dextrose/L-Lysine volumetric ratio of 0.3/1.0 (LSCO-2) after performance (L-Lysine/glucose volumetric ratio of 1.0/0.3) 14.3 $m^2\ g^{-1}$ is much larger than that of nanoparticle LSCO-4 (4.4 $m^2\ g^{-1}$). Under the conditions of 40 ml/min (GHSV = 30,000 ml/(h g_{cat}), $CH_4/O_2/N_2$ = 10/10/80 ml/min;), the LSCO-2 catalyst showed the high performance, giving the $T_{10\%}$, $T_{50\%}$, and $T_{90\%}$ of ca. 249, 461, and 702 °C, respectively. It is summarized that high O_2 adspecies concentration and excellent low-temperature reducibility were responsible for the best catalytic activity of the LSCO-2 catalyst.
2. As comparison 1D catalyst, we successfully synthesized rhombohedrally crystallized three-dimensionally ordered macroporous $La_{0.6}Sr_{0.4}MnO_3$ (3DOM LSMO) by the surfactant-mediated colloidal crystal PMMA-templating route. The LSMO-DP3 catalyst with a surface area 42.1 m^2/g performed the best, success the $T_{10\%}$, $T_{50\%}$, and $T_{90\%}$ of 437, 566, and 661 °C, respectively. They were much lower by 125, 106, and 88 °C than those ($T_{10\%}$ = 562 °C, $T_{50\%}$ = 672 °C, and $T_{90\%}$ = 749 °C) obtained on the 1 Dimensional LSMO catalyst with a BET of 2.6 m^2/g, respectively.
3. In addition, the DMOTEG/PEG400 volumetric ratio had an influence on the crystalline phase, crystallite size, and crystallinity of 3DOM LSMO. There was the copresence of surface Mn^{4+} and Mn^{3+} species on LSMO, and the surface Mn^{4+}/Mn^{3+} molar ratios of LSMO-DP3 and LSMO-DP1 were much higher than those of 1D LSMO catalyst.
4. A number of rhombohedrally crystallized 3DOM-structured $La_{0.6}Sr_{0.4}MnO_3$-supported Ag NP catalysts with high surface areas (38.2–42.7 m^2/g) were firstly successfully synthesized by a facile novel reduction method in a DMOTEG

H. Arandiyan, *Methane Combustion over Lanthanum-based Perovskite Mixed Oxides*, Springer Theses, DOI 10.1007/978-3-662-46991-0_6

solution using PMMA colloidal crystal as template. The *y*Ag/3DOM $La_{0.6}Sr_{0.4}MnO_3$ (*y* = 0, 1.57, 3.63, and 5.71 wt%) possessed unique nanovoid-walled 3DOM architecture. Among the *y*Ag/3DOM $La_{0.6}Sr_{0.4}MnO_3$ samples, 3.63 wt% Ag/3DOM $La_{0.6}Sr_{0.4}MnO_3$ possessed the highest O_{ads} concentration and the best low-temperature reducibility. The 3.63 wt% Ag/3DOM $La_{0.6}Sr_{0.4}MnO_3$ catalyst performed the best, giving the $T_{10\%}$, $T_{50\%}$, and $T_{90\%}$ of 361, 454, and 524 °C at GHSV = ca. 30,000 ml/(g h), respectively, which were much lower than those achieved over the 1 Dimensional $La_{0.6}Sr_{0.4}MnO_3$ sample by 201, 216, and 225 °C, respectively.

5. The addition of 3.0 vol% water vapor did not affect critically the catalytic activity of the 3.63 wt% Ag/3DOM $La_{0.6}Sr_{0.4}MnO_3$ sample at temperatures above 600 °C, but it decreased the catalytic activity at temperatures below 550 °C. There was a small negative effect of sulfur dioxide on the catalytic activity of the 3.63 wt% Ag/3DOM $La_{0.6}Sr_{0.4}MnO_3$ sample.
6. The 3.63 wt% Ag/3DOM $La_{0.6}Sr_{0.4}MnO_3$ sample exhibited the highest TOF_{Ag} values of 1.86×10^{-5} mol/(mol_{Ag} s) at 300 °C and 11.8×10^{-5} mol/(mol_{Ag} s) at 400 °C. The apparent activation energy (E_a) values (91.4 and 56.6 kJ/mol, respectively) of the 1D $La_{0.6}Sr_{0.4}MnO_3$ and 3DOM $La_{0.6}Sr_{0.4}MnO_3$ catalysts were much higher than those (37.5 kJ/mol) of the 3.63 wt% Ag/3DOM $La_{0.6}Sr_{0.4}MnO_3$ catalyst. The 3.63 wt% Ag/3DOM $La_{0.6}Sr_{0.4}MnO_3$ catalyst shows super catalytic activity for methane combustion, which is attributed to a higher oxygen adspecies amount, larger surface area, better low-temperature reducibility, and unique nanovoid-walled 3DOM structure.

6.2 Recommendations for Future Works

In this thesis, the applicability of nanostructure perovskite mixed oxide was studied. However, highly dispersed Ag nanoparticles supported on high-surface-area 3DOM $La_{0.6}Sr_{0.4}MnO_3$ were successfully generated via the dimethoxytetraethylene glycol-assisted gas bubbling reduction route and the macroporous materials showed super catalytic performance for CH_4 combustion, but there are still some deficiencies that need to be improved in future research work as follows:

1. The variation in CH_4/O_2 ratio can be extended for further investigations on activity/selectivity. In the meantime, mechanistic studies and theoretical modeling of reaction kinetics can be carried out in order to improve the fundamental understanding of the catalytic cycle.
2. The rhombohedrally crystallized used 3DOM $La_{0.6}Sr_{0.4}MnO_3$ (LSMO) catalysts with PMMA hard template strategy, but this thesis is not in-depth study on particle morphology structure catalyst and their relationships between its catalytic properties. It is also interesting to study on materials for long-term thermal and hydrothermal stability would be useful to ascertain their applicability in industrial application.

3. We reported the characterization, preparation, and performance activity of CH_4 combustion of 3DOM LSMO with nanovoid-like or mesoporous skeletons by the use of the surfactant [DMOTEG, L-lysine, poly(ethylene glycol) (PEG400), triblock copolymer (Pluronic P-123), and/or ethylene glycol (EG)]-assisted PMMA-templating strategy, but recommend developing this idea on fully mesoporous structure (nanovoid). Additionally, increase BET surface area by explores more innovative preparation strategies for 3DOM architecture of mesoporous perovskite-type oxide catalysts has been studied. In addition, in terms of loading noble metal NPs, there is no study on loading other precious metals or transition metal catalysts. In future work needs to porous transition metal oxides or complex oxide supported catalyst or other precious metals and transition metal oxide for CH_4 combustion reaction research.

Appendix

See Figs. A.1, A.2, A.3, A.4, A.5, A.6, A.7, A.8, A.9, A.10, A.11, A.12, A.13, A.14, A.15, A.16 and A.17.

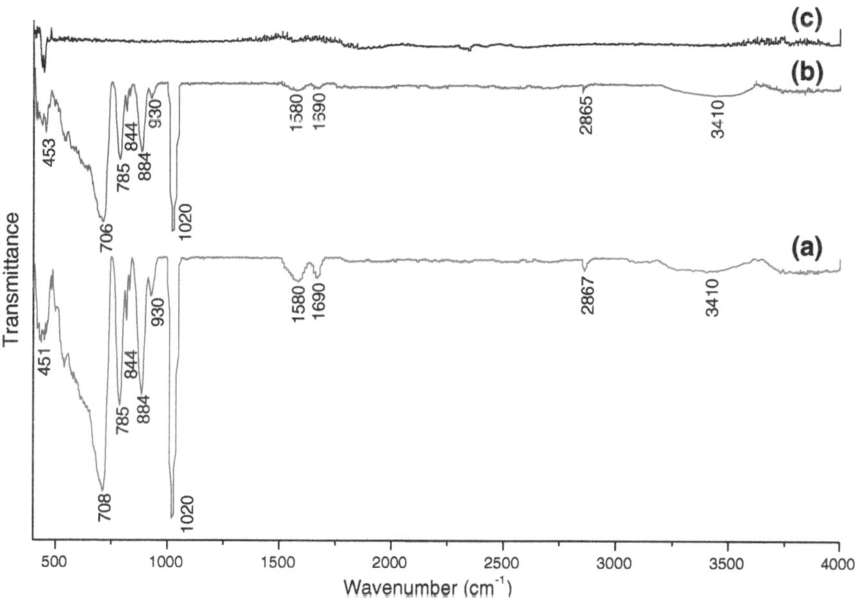

Fig. A.1 FT-IR spectra of **a** uncalcined LSMO-DP3, **b** uncalcined LSMO-DP1, **c** calcined LSMO-DP3 in air at 800 °C for 4 h

H. Arandiyan, *Methane Combustion over Lanthanum-based Perovskite Mixed Oxides*, Springer Theses, DOI 10.1007/978-3-662-46991-0

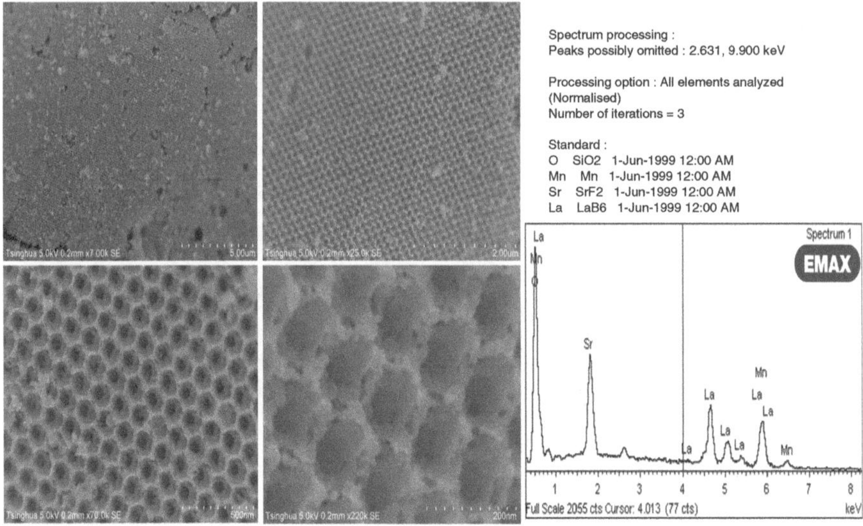

Fig. A.2 SEM images and EDS result of 3DOM LSMO-LP1

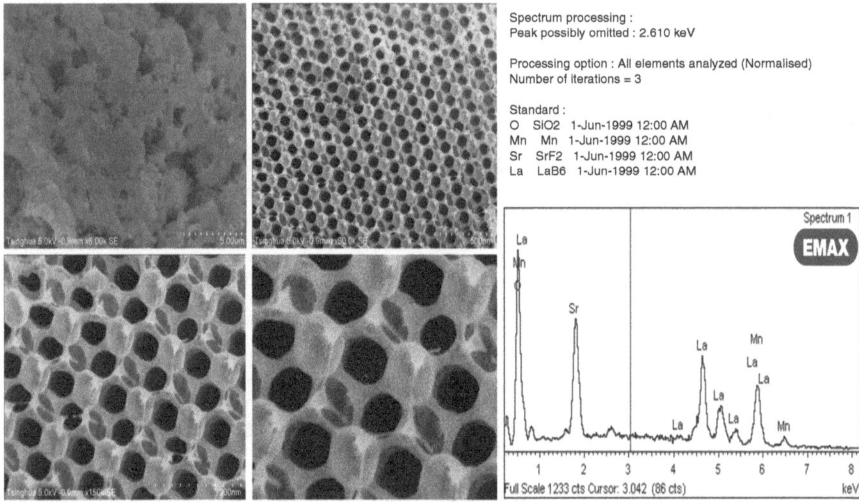

Fig. A.3 SEM images and EDS result of 3DOM LSMO-DP5

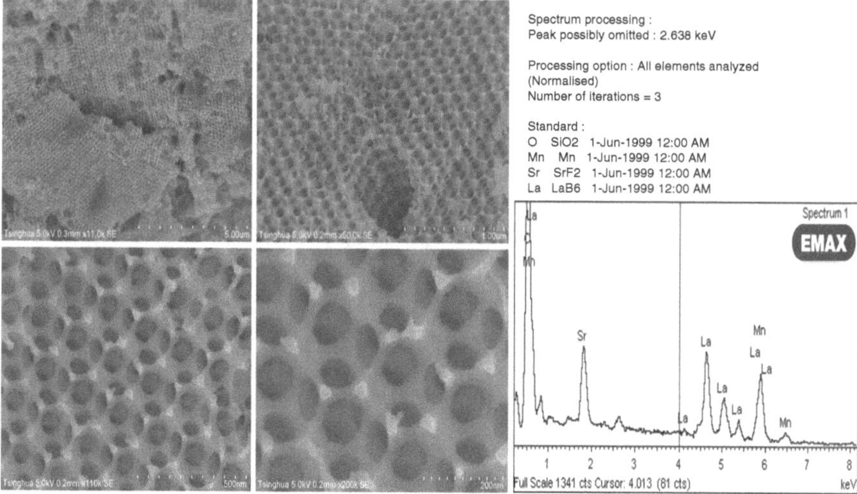

Fig. A.4 SEM images and EDS result of 3DOM LSMO-DP3

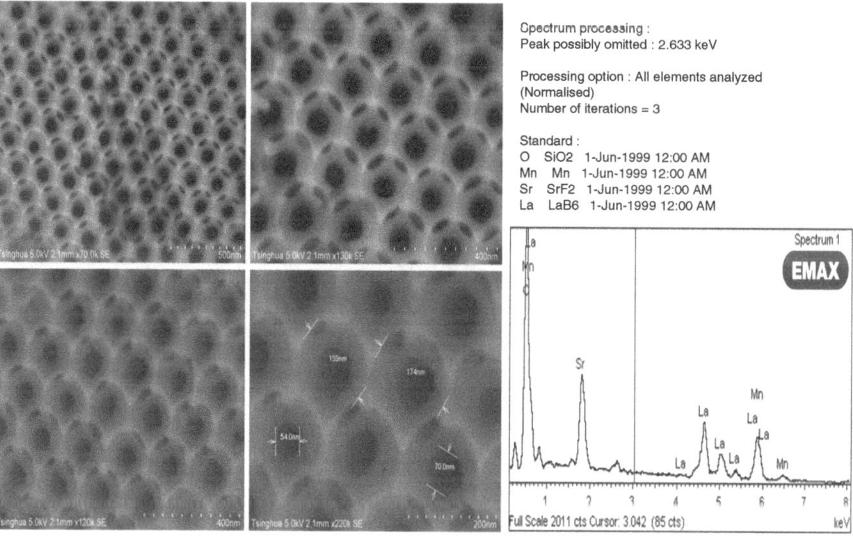

Fig. A.5 SEM images and EDS result of 3DOM LSMO-DP3-850

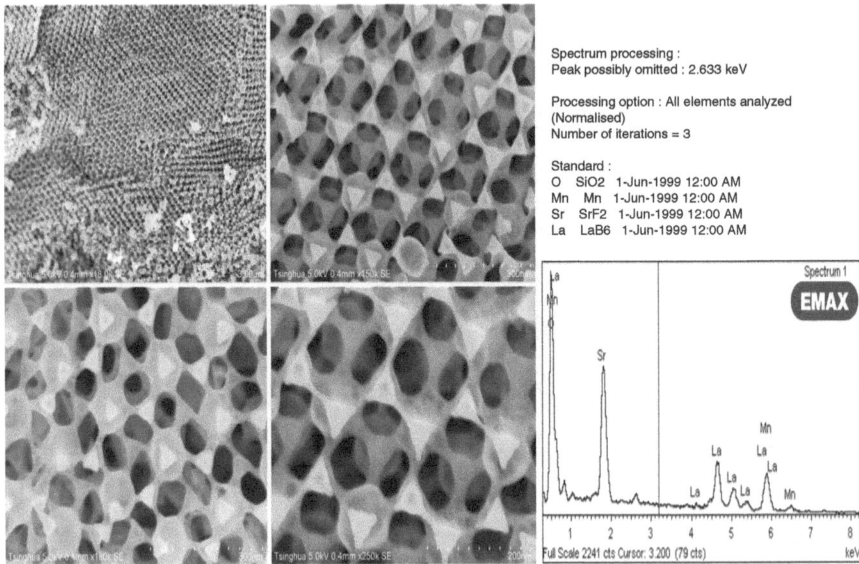

Fig. A.6 SEM images and EDS result of 3DOM LSMO-DP1

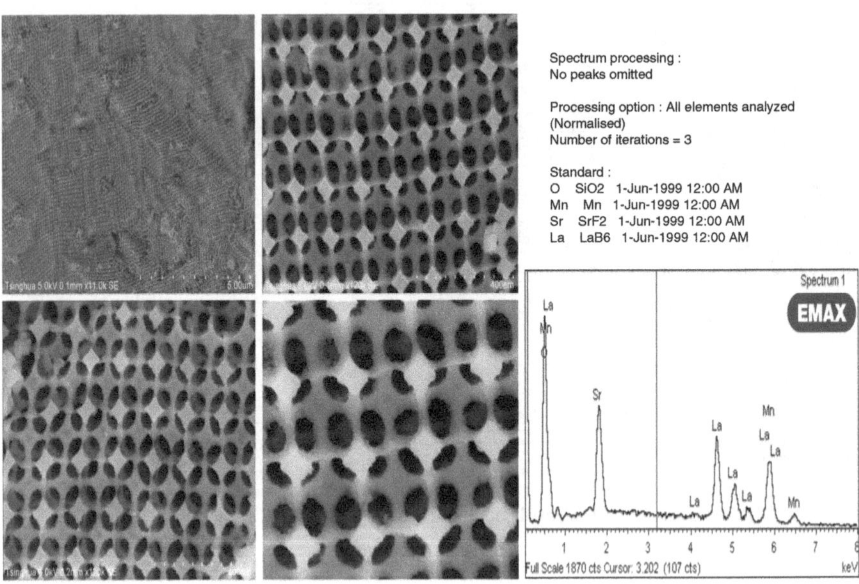

Fig. A.7 SEM images and EDS result of 3DOM LSMO-PE1

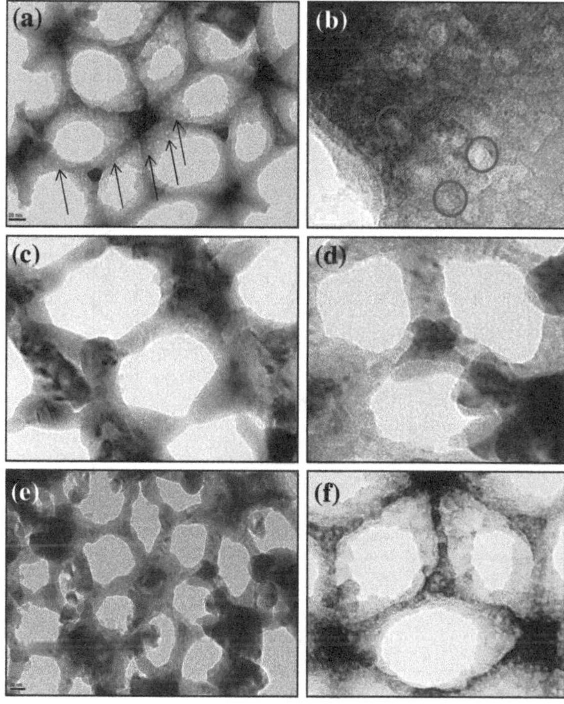

Fig. A.8 TEM images of **a**, **b** LSMO-DP3, **c** LSMO-LP1, **d**, **e** LSMO-DP5, and **f**, **g** LSMO-PE1

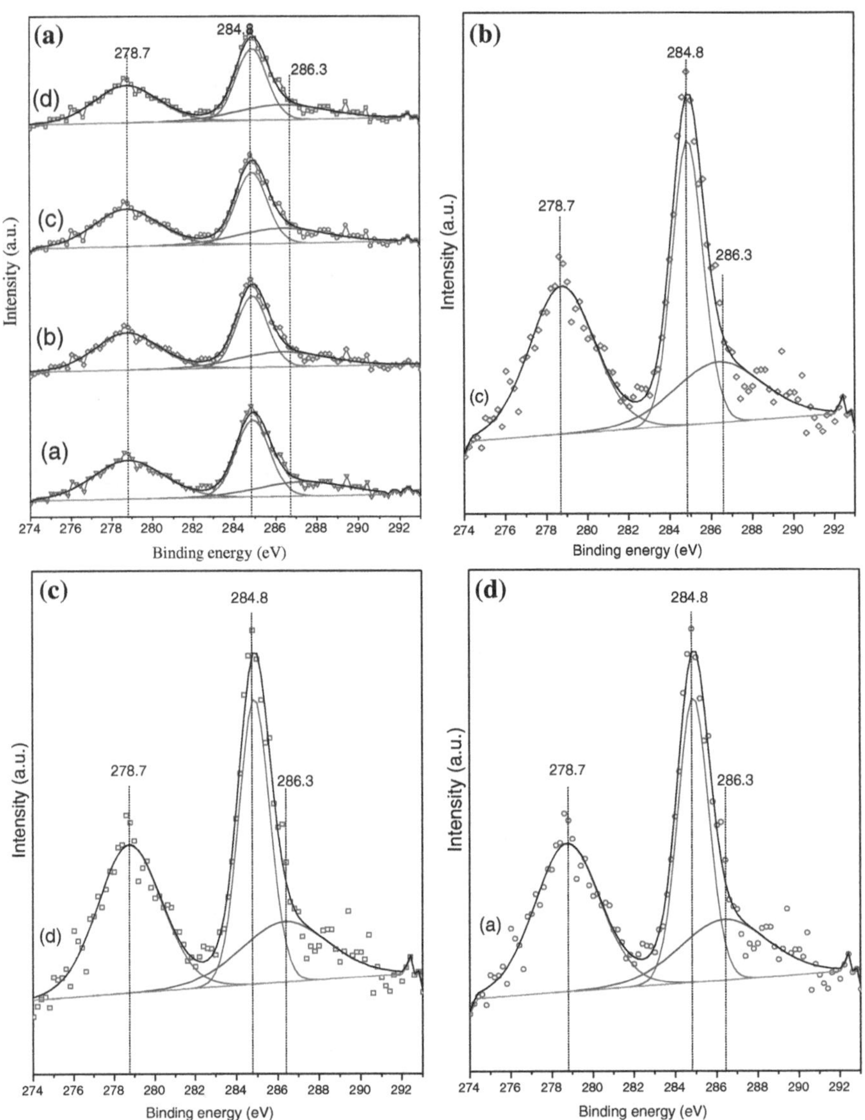

Fig. A.9 C 1s XPS spectra of **a** LSMO-DP3, **b** LSMO-LP1, **c** LSMO-PE1, **d** LSMO-PE1

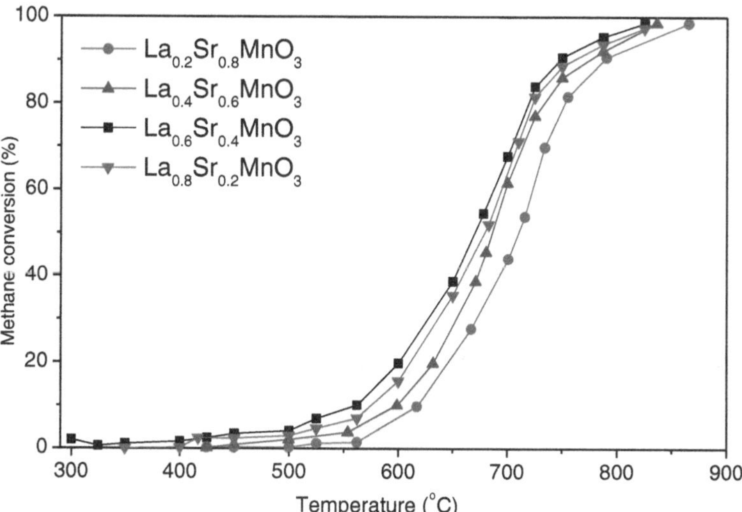

Fig. A.10 Effect of Sr substitution fraction (x) on the activity of $La_{1-x}Sr_xMnO_3$ ($x = 0.2, 0.4, 0.6, 0.8$) catalysts for the conversion of CH_4

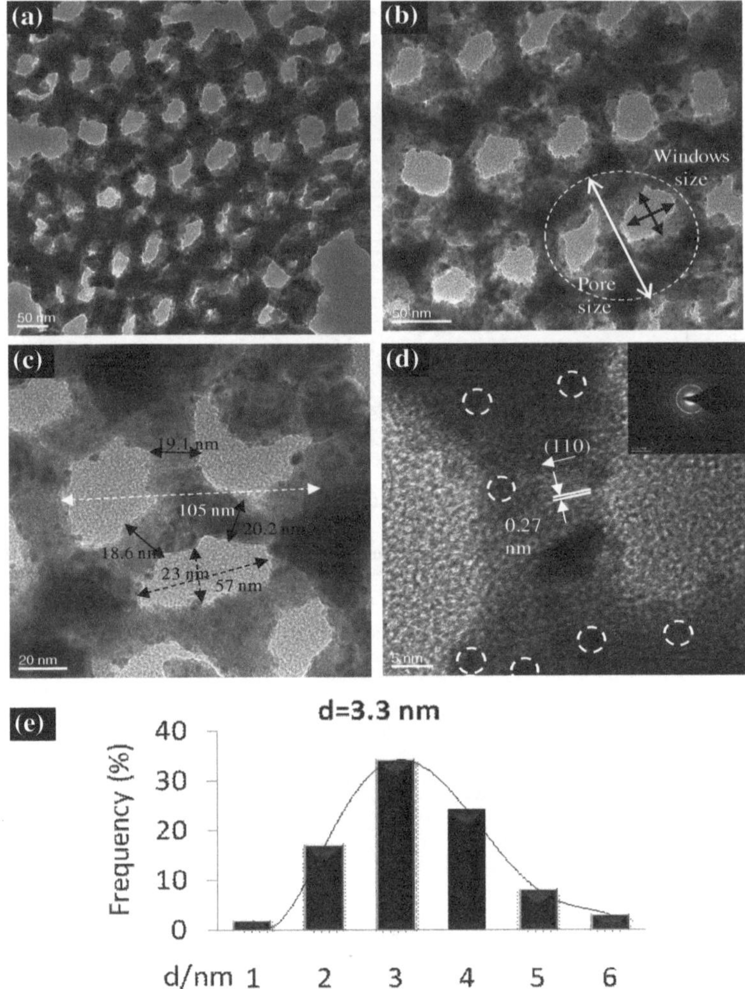

Fig. A.11 TEM images **a**, **b** and HRTEM images **c**, **d** and size distribution of Ag nanoparticles **e** of 1.58 wt% Ag/3DOM $La_{0.6}Sr_{0.4}MnO_3$. The *white circles* of **d** clearly showing the lattice fringes of Ag suggests the formation of small Ag NPs on the 3DOM $La_{0.6}Sr_{0.4}MnO_3$ support

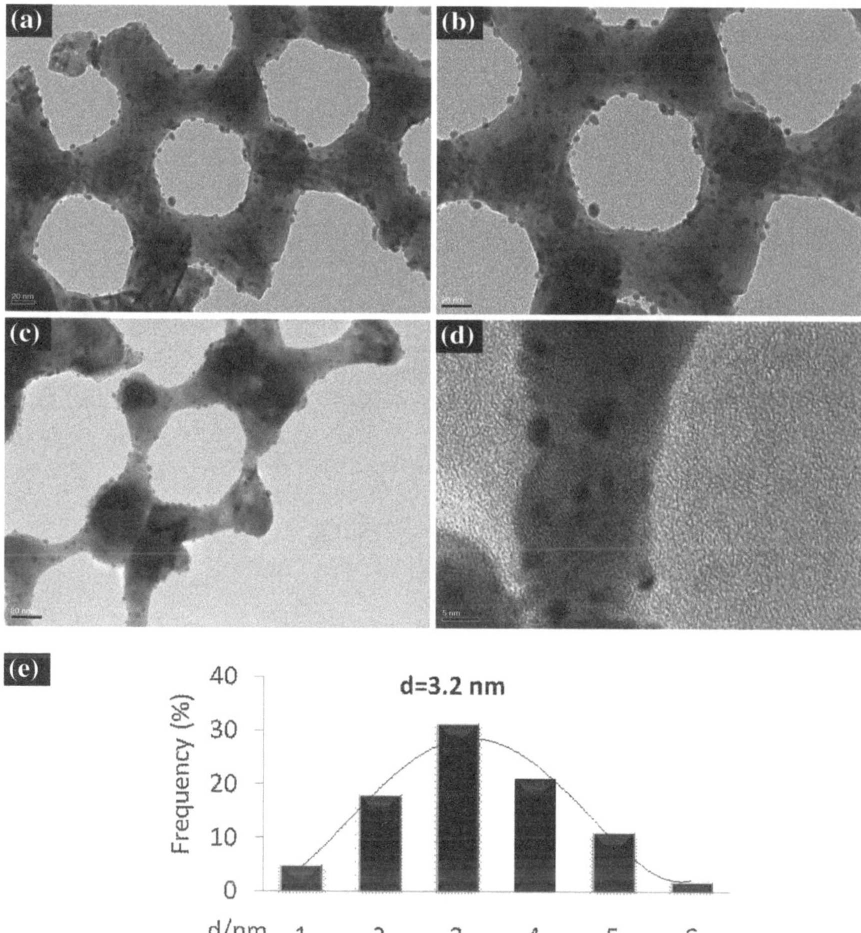

Fig. A.12 TEM **a**–**c** and HRTEM **d** images and size distribution of Ag NPs by statistic analysis of more than 200 Ag particles in the HRTEM image **e** of 3.63 wt% Ag/3DOM $La_{0.6}Sr_{0.4}MnO_3$

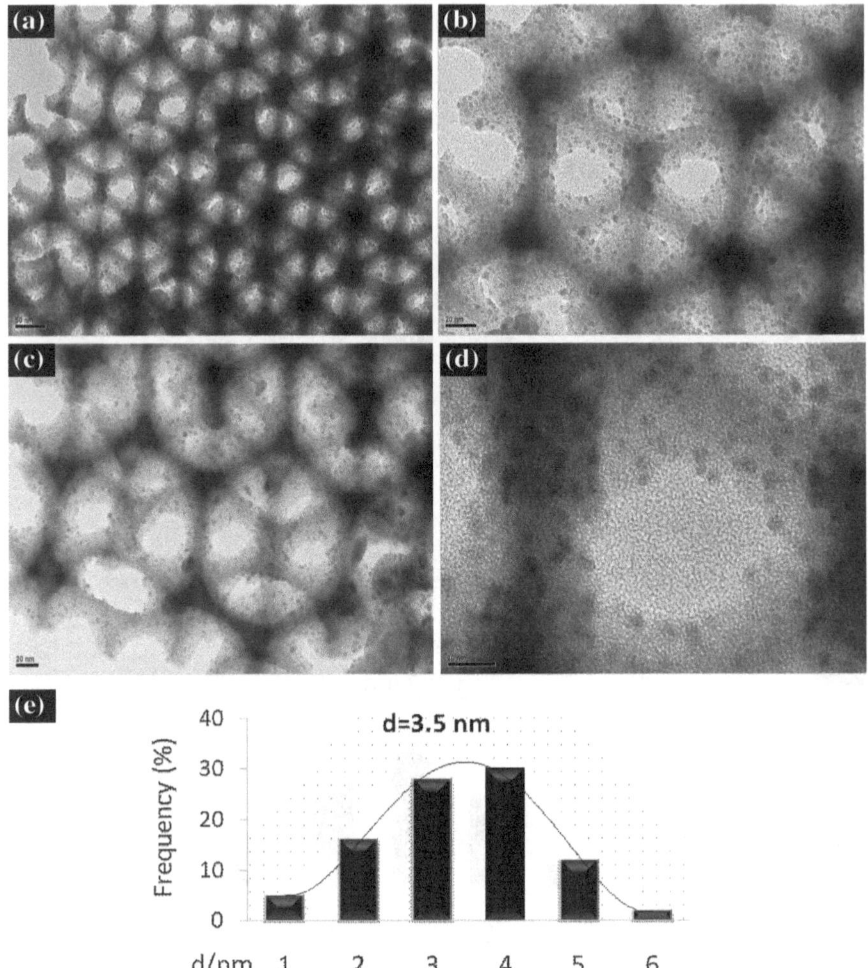

Fig. A.13 TEM **a–c** and HRTEM **d** images and size distribution of Ag NPs by statistic analysis of more than 200 Ag particles in the HRTEM image **e** of 5.71 wt% Ag/3DOM $La_{0.6}Sr_{0.4}MnO_3$ catalyst

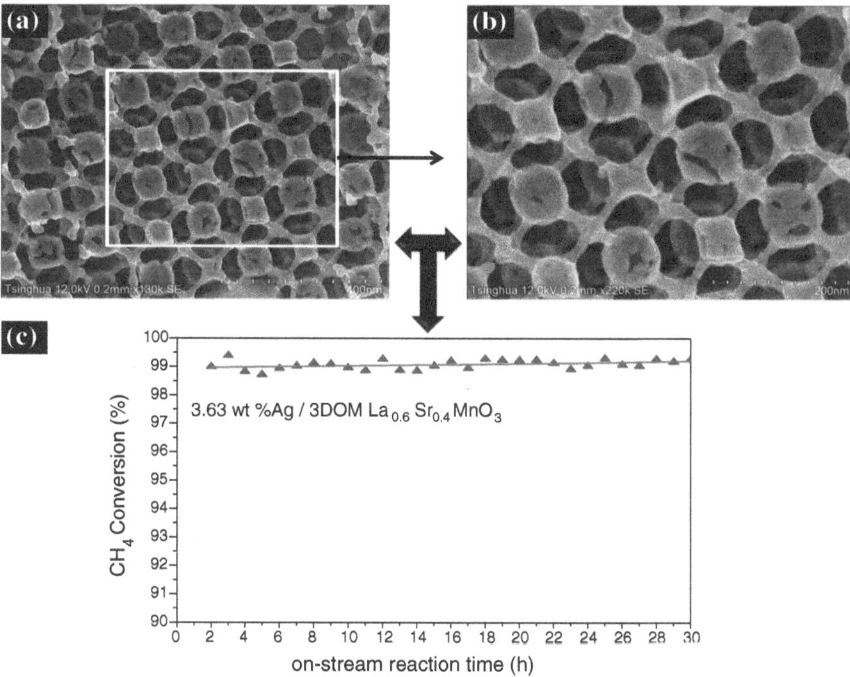

Fig. A.14 SEM images **a**, **b** of 3.63 wt% Ag/3DOM $La_{0.6}Sr_{0.4}MnO_3$ catalyst after 30 h of on-stream reaction, and **c** CH_4 conversion versus on-stream reaction time over the 3.63 wt% Ag/ 3DOM $La_{0.6}Sr_{0.4}MnO_3$ at GHSV = 30,000 ml/(g h) under the conditions of 2 % CH_4 + 20 % O_2 + 78 % N_2 (balance) and the total flow = 41.6 ml/min

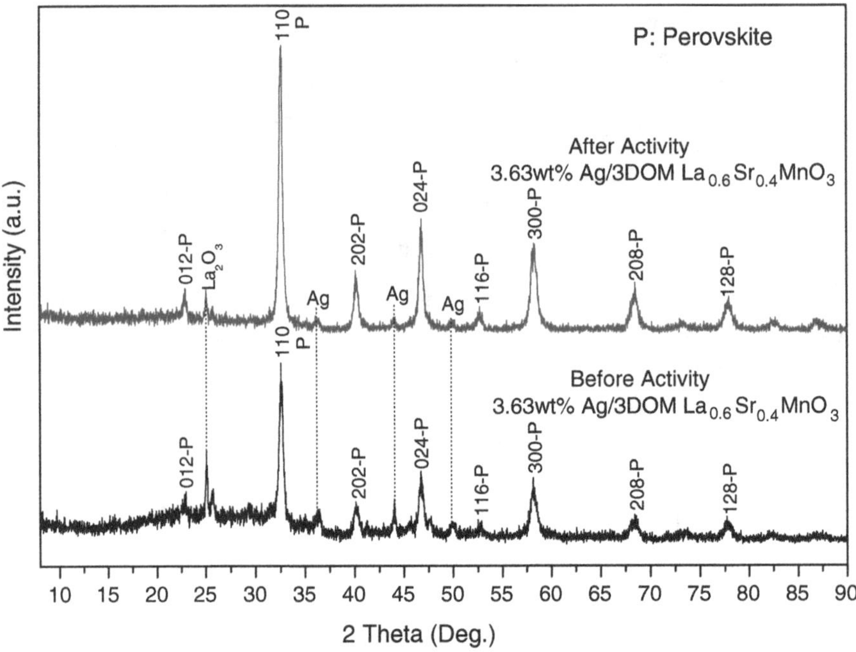

Fig. A.15 XRD pattern of the 3.63 wt% Ag/3DOM $La_{0.6}Sr_{0.4}MnO_3$ sample after 20 h of on-stream reaction at GHSV = 30,000 ml/(g h) under the conditions of 2 % CH_4 + 20 % O_2 + 78 % N_2 (balance) and the total flow = 41.6 ml/min

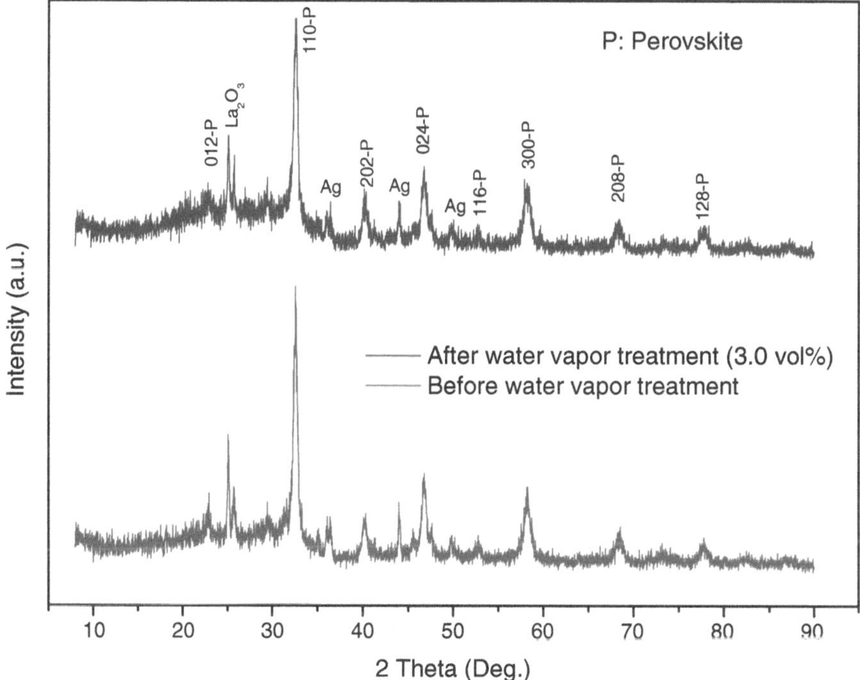

Fig. A.16 XRD patterns of the fresh and used 3.63 wt% Ag/3DOM $La_{0.6}Sr_{0.4}MnO_3$ sample after 10 h of on-stream reaction in the presence of 3.0 vol% H_2O under the conditions of GHSV = 30,000 ml/(g h) under the conditions of 2 % CH_4 + 20 % O_2 + 78 % N_2 (balance) and the total flow = 41.6 ml/min

Fig. A.17 **a** HRSEM and **b** HRTEM images of the used 3.63 wt% Ag/3DOM $La_{0.6}Sr_{0.4}MnO_3$ samples after 10 h of on-stream reaction in the presence of 3.0 vol% H_2O

Zeitfracht Medien GmbH
Ferdinand-Jühlke-Straße 7
99095 Erfurt, Deutschland
produktsicherheit@kolibri360.de